FOREWORD

There is little published material available in the UK on the incentivisation of contracts. This factor coincided with CIRIA's response to the increasing concern within the construction industry of the pressure on pricing using traditional tendering processes, with the consequent impact on the time and quality aspects of projects. The rise of the "claims culture" had led to conflict between clients and contractors, with "winners" and "losers" emerging. A study on the incentivisation of contracts was not intended to seek a panacea but rather to provide an alternative procurement route when more traditional methods were likely to create potential conflicts – with subsequent detrimental effects on the overall success of the project.

The key objectives upon which the study was established were:

- to examine the need for, and the implementation of, construction contract incentive schemes, and to illustrate recent innovations in particular
- to analyse and compare the success or otherwise of the schemes, and to relate this to the various forms of construction contract and procurement route
- to publish a practical and authoritative report aimed at all participants in the construction process.

This approach was adopted to support the premise that positive incentives are a consideration for ensuring successful contracts.

The success or failure of major projects in the construction industry is generally measured by the achievement or otherwise of the principal contract parameters of time, cost and quality. The use of traditional forms of contract have frequently led to confrontation between client, consultant and contractor where, due to many reasons, projects have incurred serious cost overruns and delays. Many industry observers have commented that there had to be a better way of procuring and completing a construction project.

Whilst all matters connected with achieving quality are seen as a common objective by all parties to the contract, being able to fix time and cost with some certainty remains the central issue. To quote from the executive summary of Sir Michael Latham's report, *Constructing the team* (1994):

> *"endlessly refining existing conditions of contract will not solve adversarial problems. A set of basic principles is required on which modern contracts can be based"*.

ACKNOWLEDGEMENTS

A contract to carry out the research and write the report was awarded to Mouchel Consulting Limited. The report was drafted by Messrs C Fuller, P Jobling and G Mcleary and edited by Mr D Richmond-Coggan. The project was overseen by a widely representative Steering Group, comprising:

Mr T P Gorman (chair)	Amey Rail
Mr R Bishop	Laing Technology
Mr J Brown	Sir William Halcrow & Partners
Mr B Butterworth	Taylor Woodrow Construction (Southern)
Mr T J Carr	British Nuclear Fuels
Dr W Hughes	University of Reading
Mr G Jamieson	Highways Agency
Mr I N L Jones	Water Service (NI)
Mr P S Knight	London Underground
Mr R H Lupton	Franklin & Andrews
Mr G Phillips	West of Scotland Water
Mr N Proctor	Brown and Root Limited
Mr D Rogers	Southern Water Technology Group
Mr J Temprell	Henry Boot Construction (UK)
Mr J H O Williams	Yorkshire Water.

Corresponding member:

Mr G Heald	Environment Agency (Anglian Region).

CIRIA's research manager for the project was Mr R Bishop, succeeded by Mr G M Gray.

The project was funded by the Highways Agency, Water Service (NI), Scottish Water Authority and CIRIA's Core Programme members. CIRIA are grateful for help given to this project not only by the members of the Steering Group and the corresponding members, but also the many individuals and organisations who were consulted in the course of the field research. In accordance with undertakings given to the latter, these contributors are not identified.

The results of the study were presented at a Construction Productivity Network workshop in May 2000. A list of the organisations attending and a report of the meeting are shown in Appendix B. Some of the issues raised at the meeting have been incorporated into this report.

CIRIA C554

London, 2001

Construction contract incentive schemes –
lessons from experience

Prepared under contract to CIRIA by Mouchel

D Richmond-Coggan MA CEng MICE

CIRIA *sharing knowledge ■ building best practice*

6 Storey's Gate, Westminster, London SW1P 3AU
TELEPHONE 020 7222 8891 FAX 020 7222 1708
EMAIL enquiries@ciria.org.uk
WEBSITE www.ciria.org.uk

SUMMARY

This publication summarises the findings from a survey of 20 construction projects which used incentivised contracts. Clients, contractors and consultants from the private and public sector were interviewed and the detailed case studies form part of the report. The studies covered utilities, transportation, civils infrastructure, building, and heavy and light process manufacturing. The report interacts with recent important guidance documents aimed at the construction industry, such as the Latham and Egan Reports, and the guides published by the Office of Government Commerce.

The "what", "how" and "when" issues relating to incentive schemes are discussed and conclusions are drawn from both the case studies and a Construction Productivity Network workshop.

Construction contract incentive schemes – lessons from experience

Construction Industry Research and Information Association

Richmond-Coggan, D

© CIRIA 2001 C554 ISBN 0 86017 554 5

Keywords		
Construction management, procurement, project management, supply chain management		
Reader interest	**Classification**	
Construction clients, clients' advisors, contractors	AVAILABILITY CONTENT STATUS USER	Unrestricted Advice/guidance Committee-guided Construction clients, clients' advisors, contractors

CONTENTS

LIST OF TABLES

LIST OF FIGURES

EXECUTIVE SUMMARY

Successful contracts cannot be guaranteed. However key factors exist which, if properly addressed, can minimise the likelihood of one or other of the parties to the contract feeling aggrieved with the outcome. The old adage that clients should look after the interests of contractors and that contractors should look after the interests of clients rings true when considering incentivisation of the contract as a procurement strategy.

This study has drawn on the incentivisation of contract experiences from a broad cross-section of sectors covering utilities, transportation, civils infrastructure, building, heavy and light process manufacturing using a series of case studies. It has provided research findings from which conclusions have been drawn and recommendations made to assist potential users.

The findings result from a survey of experiences of involvement with incentivised contracts on 20 projects within the construction industry. Over 50 interviews were held with clients, consultants and contractors associated with these projects to determine their experiences of using the incentivised approach.

In parallel with this research, a desk study of current literature was undertaken to ascertain what reference material was available to guide potential users.

Since the case study reviews were carried out, a key report on performance within the construction industry, *Rethinking construction* by the Construction Task Force, chaired by Sir John Egan, has been published. This reinforces the results of our findings that there are key drivers within the incentivisation process which, when properly focussed and controlled, can lead to better performance and hence increase the overall value to all parties. The Egan Report stated:

> "*clients need better value from their projects and construction companies need reasonable profits to assure their long-term future. Both points of view increasingly recognise that not only is there plenty of scope to improve, but they also have a powerful mutual interest in doing so.*"

MAIN FINDINGS

- **Commitment to apply incentivisation** – most of the parties to the projects (15 out of 20) positively committed themselves to the development and application of the incentivisation process. Of the remainder, the outcome of the project (either a partial success or a failure) was largely as a result of the less-than-positive commitment in putting in place a management framework to allow the scheme to work. Generally, the greater the awareness of the risks and issues encountered during the tendering process, the easier it was to include a more structured approach to managing such risks and issues within the contract. This may explain why there is a correlation between the initial and continuing commitment of the parties to the incentive scheme and the successful outcome of the project. A key factor in successfully applying incentivisation to the contract is the emphasis on selecting the right contractor. This is of equal importance to the type of incentive scheme adopted. In addition there is a case for the client site staff and their consultants being included in the incentive scheme, so they benefit from delivery as well as the contractor.

- **Contract strategy** – 17 of the schemes had a well-defined contract strategy to enable each scheme to operate. Of the remaining three schemes, two schemes were initiated as a result of reactive proposals to recover failing projects and the contracts required amendment to handle the changed circumstances. This proved successful on one project and on the other had little or no effect due to unrealistic time deadlines. The third scheme related to the semi-conductor/manufact-uring facility, which is typical of the high value product sector where informal agreements are sufficient to encourage the contractor to meet time deadlines, since the financial cost to the client is relatively low compared to the potential sales income of the product.

- **Development and administration of the scheme** – when incentivisation was included in the project, the form of contract did not appear to deter the contractors, when tendering for the work. There was however little deviation from using a simple target savings approach with open book accounting. The utilities sector appears to have favoured open book accounting to reach a target cost in developing their incentive schemes. A key issue on all such schemes was the higher level of administrative effort required to operate them. This was in part due to the lack of client experience in operating such schemes and in developing the level of trust needed to meet the common objectives. The partially successful or failed projects (seven of the 20) were generally as a result of poorly developed schemes. To ensure that contractors have the fullest understanding, the tender documents must include a full explanation of the client's objectives and values. Properly prepared documentation is essential and should not constrain either party and informal arrangements may still apply.

- **Overall outcome and measurable benefits** – of the 20 projects, 13 were deemed successful, where time and cost targets were met; five were a partial success, where either time or cost targets were achieved and the two failures where neither time nor cost targets were met. Overall, quality was found not to be an incentive issue since this was a prerequisite for any payments to be made by the client whether the works were on time or not. Most of the seven partially successful or unsuccessful projects involved time overruns. This has led in a few cases to the application of LADs and litigation.

- **The attitudes of the parties involved** – this finding relates to the continuing commitment of the parties involved in managing the contract. The majority of parties to the projects (15 of the 20) remained committed to the process. The continuity of key personnel on a project was considered a significant contribution to the eventual success of the project. Where considerable problems had arisen the incentive scheme encouraged their resolution on a collaborative basis.

- **Problems and issues which need to be resolved for future projects of a similar nature** – the most significant area of concern when considering all incentive schemes is the level of risk assessment undertaken. In 12 of the 20 projects examined, sufficient detail of the potential risks was provided at the tendering or negotiation stages for the impact to be minimal. The problems that arose on the remaining eight projects were largely as a result of a lack of investment in financial and monetary terms in determining the likely affect of such risks on the time and cost targets. Where clients had expended considerable effort in reviewing risk, it was found to be a major contributor to the successful operation of the scheme and the eventual outturn of the project. Overall, clients were of the opinion that risk assessment should be the precursor to incentivisation in the procurement process.

- **Other responses which may assist potential users in determining whether to use an incentivisation approach** – the successful schemes were achieved through a combination of motivational and commercial objectives and it is worthwhile reiterating those which should be applied in every case. They are not listed in any particular order and the relevance of each factor will vary.

  - The wholehearted commitment by all parties to achieving the project objectives.

  - A comprehensive understanding, obtained during the tendering process, of the likely risks which may occur.

  - The development of trusting relationships between the parties.

  - The alignment of the financial and commercial objectives of each party and the willingness to adopt, if necessary, a change in culture to make this happen.

  - The development of contract documentation with potential suppliers and contractors before tendering contributed to the principle of "no surprises".

  - The use of model forms of contract, developed by the specific industry sector through representative bodies, to encourage greater participation in the approach.

  - To apply best practice, financial controls must be put in place to monitor the performance of clients and contractors. These controls can be executed by the client at one extreme and by the contractor at the other extreme, or jointly.

MAIN CONCLUSIONS

The parties involved with incentive schemes considered that when properly set up, improved performance was achieved and that significant benefits accrued to all parties. Allied with the challenge to traditional forms of procurement, high spending organisations can be expected to plan future capital programmes more effectively knowing that the likelihood of achieving the completion of projects within time and cost, and to the required quality, is considerably improved when adopting an incentivised approach. Some key points from the research and from the Construction Productivity Network workshop support this statement.

- **Trust** – this is a key factor to success. The need for trust coupled with openness in all dealings and transactions cannot be too highly stressed. The time pressures on existing procurement methods are demanding new types of business relationships based on trust.

- **Potential benefits** – these are both "hard" and "soft". Regarding the latter (which could be outside any contract) the potential benefits may not always be obvious and consequently need to be communicated to those lower in the organisation. They may also be difficult to justify commercially.

- **Incentive schemes require a higher level of administrative effort** – any additional administration costs should be more than outweighed by the benefits derived from the incentive scheme.

- **Risk** – a detailed risk review is an essential precursor to any incentive scheme. It should only be undertaken by those with the appropriate knowledge and skills.

- **Incentive schemes need to be well documented and communicated to all parties to the contract** – ensure that the parties (client, contractor and other consultants) understand the incentive scheme and that any potential "blockers" to it are identified. During the project if it becomes necessary to address issues there is advantage in attempting to resolve these initially at the lowest management level. In allowing issues to rise up within an organisation, the opportunity for early resolution may be lost.

- **The development and implementation of incentive schemes may require the acquisition of new skills** – these will especially be required by those connected with culture change and interpersonal skills, such as communication. The latter is particularly important both within and between each party to the contract and from the highest to the lowest levels of the personnel involved. The client's ability to communicate with the contractor is a key determinant to success.

- **Safety** – this should not be compromised by any incentive scheme. Costs related to safety should also be included, eg for training. Safety is likely to be considered together with the environmental issues relevant to the project. Improvements to safety and enhancement of the environment through better performance will add significant benefits.

- **Contract incentive schemes are not necessarily an easy or comfortable option and need genuine commitment from all concerned in "a different way of doing business"** – there are greater opportunities to benefit from incentive schemes applied to rolling programmes of work on the basis of continuous improvement. However, it is vital that the parties do not become complacent and that there is no drop in performance.

- **Incentive schemes should be seen as aiming to drive down inefficiencies and costs and not attacking the contractor's margins** – the contractor is looking for profit, repeat business and industry recognition. It is essential to understand the cost base so that the right element(s) (who and what) is incentivised.

The evidence obtained from the parties suggests that most clients would seek to repeat the process on new projects but with greater commitment to reviewing risk during the tendering process.

During the period of recession in the 1990s, contractors may have taken on more risk than they would normally do, just to win contracts. The evidence from clients would suggest that they are acutely aware that if too much risk is passed to the contractor, there is a greater chance that, if things go wrong, they could be faced with the potentially high cost and effort to fight claims. Clients are not guaranteed of success and the impact on budgets could be catastrophic. The opportunity to minimise the adversarial attitudes generated by construction contracts and gain some mutual benefit as to time and cost is certainly an avenue most clients would explore.

Incentive schemes that are set up during the pre-contract stage are likely to be less problematic since the procurement process should ensure a "level playing field" for all tenderers and there is an audit trail inherent in the process. Where the parties had invested more time during the tendering process to align their objectives and assess the risks, then the indications are that the chances of a successful outcome remain high. Where there remains a degree of scepticism or even some mistrust, then the potential benefits may not be fully realised in operating the scheme.

The most successful schemes emanate from total commitment and trust by all parties working towards a common aim. This is largely brought about by a combination of two major factors.

1. The investment by the client in analysing project risks in sufficient depth at the tender stage, coupled with an encouragement to potential tenderers to innovate and look at whole life costs.

2. All parties are involved in creating the right environment where accountability and responsibility is delegated to the level where problems could be resolved and solutions implemented. The workings of each scheme dictate a requirement for close harmony and from this set of circumstances, expectations can be matched by achievements.

THE WAY FORWARD

This study suggests, through the experiences of the parties participating in the survey, that implementing an incentivisation scheme can be an effective way of ensuring that the desired outcome, in terms of time, cost and the management of risk can be achieved with some certainty, whilst still achieving the required quality.

This point is emphasised in the Egan Report (1998), which wished to see:

> *"the introduction of **performance measurement** and competition against clear targets for improvement, in terms of quality, timeliness and cost, as the principle means of sustaining and bringing discipline to the relationships between clients, project teams and their suppliers. Partnering should not be a soft option. The evidence we have seen is that these relationships, when conducted properly, are much more demanding and rewarding than those based on competitive tendering. There are important issues here for the public sector".*

In this respect a number of issues could be considered to improve the chances of success.

1. Whilst each scheme must be tailored to the particular needs and jointly developed by the parties, each sector of the construction industry should be encouraged to develop model schemes to take account of the specific issues peculiar to that sector, for example:

> Within the wastewater treatment sector, the optimisation of the treatment process is seen as a key objective which requires experience and knowledge of new plant and equipment to meet specified treated water targets. Developing a model contract form based on the experience of several utilities could provide the consistency of approach to incentivisation.

2. As part of the overall development of the use of incentivisation, all parties should commit more resource towards the change in culture and communication skills that may be needed to make the scheme work effectively, for example:

> The use of partnering concepts can provide the management and operational environment in which to develop this new culture and consider incentivisation proposals but partnering is not necessarily an element within an incentive scheme (see Section 3.1).

3. With the development of public/private partnerships as part of the best value initiative, there is the opportunity to use the negotiated procedure in a wider context with greater effect when applied to construction contract incentive schemes, for example:

> The audit trail created during the negotiation phase must be sufficiently robust to explain how each scheme is formulated and what the measurable benefits will be. This safeguard will encourage a wider consideration of incentivisation, particularly on complex projects.

SCOPE OF THE STUDY

In 1994 Sir Michael Latham published his report, *Constructing the team*, which confirmed that traditional contracts might not be meeting the needs of modern business. At that time, many construction contracts were based on standard or organisation specific contracts which were essentially straightforward to enter into and consisted of standard terms. One of the advantages of using standardised forms of contract is their ability to minimise legal costs by removing the need to negotiate specific contract terms. In addition, most professionals within the construction industry are familiar with them.

Since then some organisations (particularly the oil, gas and water industries) have taken steps to adjust their standard contracts to take account of the sort of relationship that ought to exist between clients (or promoters), their professional advisors and contractors leading to less confrontation. These amendments included more equitable risk sharing arrangements and performance management measures such as incentivisation. Incentivisation is the subject of this study.

In the construction industry, higher levels of competition and associated low pricing have led contractors to focus on maximising income. This could conflict with client objectives with regard to contract programmes and the quality of the final products(s) and longer-term procurement plans. In order to overcome some of these problems a variety of different contract "incentive schemes" have been introduced in an attempt to align the objectives of clients and their contractors.

This study has examined evidence from some incentive schemes that have been implemented in UK construction contracts and from the specific literature on the subject to deduce what lessons can be learned and passed on to those who wish to implement such schemes in the future. The emphasis of the study was placed on:

- identifying the different types of contract incentive schemes in current use in the UK
- analysing the effect of these schemes upon the attainment of project objectives
- correlating the success, or otherwise, of the schemes with particular contract types and procurement routes
- identifying particularly successful, or particularly unsuccessful, combinations of contract incentive scheme, forms of contract and procurement route, together with the degrees of influence and precedence each has upon the other.

It examined the experiences of the parties involved in some 20 contracts, covering utilities, transportation, civil engineering infrastructure, building, heavy and light process manufacturing, together with the experiences of the research contractor and considered the lessons learned. Around 50 interviews were held with client and contractor representatives to determine their experiences in using an incentivised approach. The results have been analysed to provide guidance on the use and improvement of incentives within contracts. It seeks to answer the questions that might be posed by clients (or their advisors) wishing to establish or participate in contract incentive schemes.

> The scope of the study defined and the process highlighted to show how evidence of incentivisation would be gathered to provide findings upon which to draw conclusions. The process also indicated how guidance to the reader would enable them to apply incentivisation in their particular case.

STRUCTURE OF THIS BOOK

This publication is addressed to all clients, contractors and their advisors, with the expressed intention of drawing on the experiences of those within the construction industry who have taken this route and generally been successful.

It assumes that readers already have a basic understanding of the fundamentals of English contract law and are reasonably knowledgeable of traditional procurement regimes in the UK, but may have only limited experience of incentivised contracts.

By making reference to case study material, the reader will be guided on how best to apply incentivisation in his or her own situation:

- to determine if an incentive scheme would be of value in a given set of circumstances
- to select an appropriate incentive scheme to suit the circumstances
- to set out the management of the scheme to ensure that the scheme will be successful.

There are two main parts.

1. A report of the main findings and discussion of the issues arising, together with practical advice and guidance on the application of incentivised contracts.

2. Selected interview material from the projects studied in depth, presented as case studies. They provide the specific facts and evidence for each project and can be used as a benchmark for projects of a similar nature.

The appendices follow. Appendix A comprises three helpful toolboxes and Appendix B is the full report of the Construction Productivity Network workshop of May 2000, at which the results of this study were presented.

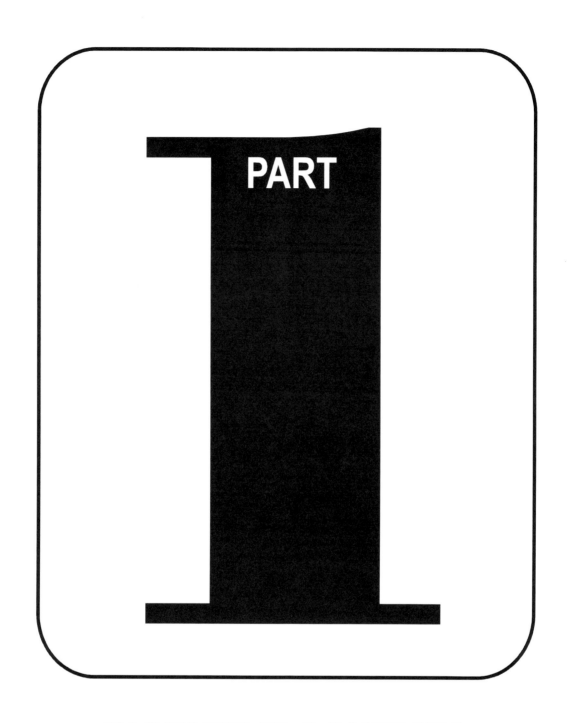

PART 1

INCENTIVE SCHEMES: THE ISSUES

1 WHAT ARE INCENTIVE SCHEMES AND WHY USE THEM?

The purpose of an incentive contract is to highlight the matters of particular concern to the client and increase the chance of overcoming them by encouraging the contractor to perform. Aligning the objectives of the client and contractor through the use of suitable performance measures and linking them to payment generally brings this about. In this way both parties share in the risks and rewards of achieving the desired result. From the client's perspective (at its most basic level) it is about moving the contractor's focus away from a short-term financial gain to looking more broadly at the long-term benefits to the client, ie life cycle costs and longer-term relationships, that consider the long-term effects on the client's business and not just the direct cost of the contract.

The broader objectives will include contract completion dates, the quality requirements and the risks to concluding a successful outcome to the contract. Therefore incentivised contracts are about delivering a performance that is "better" than that which could be achieved under a standard contract; the "better" being the greater certainty of delivering the client's desired performance through a detailed agreement on the sharing of risks and the associated rewards, for example:

> The contractor makes better use of occupation time (road/rail) by increasing productivity through applying innovative techniques.

It should be noted that the performance objectives should be realistic and achievable, otherwise the contract will probably fail, no matter how attractive the incentive(s). The Procurement Guides circulated by the Office of Government Commerce for consultation (2000) indicate that:

> *"an incentive mechanism is a process whereby parties to a contract are rewarded for performance significantly over and above that contracted for, which is of material benefit to the client and can be measured. Incentives may be financial or non-financial or a combination of the two".*

As already suggested, performance within a construction contract is conventionally measured by reference to time, cost and quality, where:

- time is the period taken for the client's construction requirement (or an identifiable element of it) to be completed compared with the construction programme agreed/set

- cost is the final sum paid for delivering the client's construction requirement (or an identifiable element of it) compared with the sum initially agreed

- quality is the reflection of the success of the completed work in complying with the requirements of the contract.

Thus "better" performance will either:

- beat the completion date
- beat the construction price
- beat the quality requirements
- improve safety and environmental requirements.

The better performance will only be achieved if you effectively manage the risks associated with the construction project to the satisfaction of all stakeholders ie clients, consultants and contractors.

Consequently, if there is a need to deliver any of these improvements, then an incentive scheme may be a suitable approach to achieve this, particularly for:

- new projects
- projects that are underway but in no difficulty, but where changes in circumstances may need a different approach
- projects that are underway but may be in difficulty.

In this respect the study identified three categories of incentive:

- pre-planned
- responsive
- reactive.

These are defined in Table 1.1. It also found that the reason for introducing the incentive scheme has a significant impact on the involvement of the parties to the contract in defining and developing the scheme.

It can be seen from the table that in both the "pre-planned" and the "responsive" categories, it is likely that the client and the contractor will develop their early incentivisation proposals in isolation from each other. However, the study identified that mutual benefit is a necessary characteristic of successful incentive schemes. Best practice is for both parties to develop incentives together by developing an understanding of the other's values, aims, objectives and constraints so that the proposed incentive scheme(s) is seen as a desirable benefit to both parties.

In the case of "pre-planned incentive schemes", this may be achieved through general consultation, possibly with recognised industry bodies during the inception phase of the project. In the case of "responsive incentive schemes", this presents more of a problem, as the contractor is likely to wish to keep its proposal confidential until it is submitted. It is likely, therefore, that in both these cases there will need to be some negotiation after submission of tenders.

In the case of "reactive schemes", the parties will already be in a contractual relationship when one of the parties foresees a mutual benefit in changing the nature of the contractual relationship by introducing incentivisation. This could result from the potential for both sides to benefit from a change in construction methodology, or from an unforeseen event leading to a need to change the balance of risks between the parties. Whatever the motivation, the study found that a high level of confidence between the parties is a key indicator towards ensuring a successful incentive scheme. Without it the relationship between the parties is likely to interfere with the open discussion of a better way forward.

After considering the potential benefits to be gained from incentive schemes in contracts, the remainder of this publication will look at how they work and how to implement them.

Table 1.1 *Incentive categories*

Category of incentive	Stage or type of project	Likely motivation	Way incentive is proposed
Pre-planned	A new project or programme of projects	Client wishes to obtain "better" performance or gain benefit from improved performance	By client before contractors are invited to tender or negotiate for the contract
Responsive	A new project	Contractor trying to win a contract by proposing an added benefit for the client	By contractor during tender
Reactive	A project that is under way	Either the client and/or contractor see a mutual benefit by proposing a change	By either party during the construction, once the opportunity becomes apparent
	A project that is underway but in some difficulties	Either the client and/or contractor trying to save a project that has run into difficulties from technical aspects	By either party during the construction, once difficulties become apparent

1.1 POTENTIAL BENEFITS

In order to gain the potential benefits of introducing an incentive scheme, it is necessary to align the business aspirations of the parties. This should consider the commercial objectives of the contractor and the desirable outcomes, as defined by the client(s).

> For example, in Case study 1 (Road bypass scheme), the client was determined not to repeat previous difficulties that occurred on an adjacent site and wished to involve the successful tenderer with a more equitable set of desirable outcomes. This included a payment mechanism that allowed the contractor's cash flow to remain positive throughout the contract period. Once the appropriate working relationships were established, the use of the incentivised approach was found to be of mutual benefit and resulted in both parties working towards "better" performance.

On complex projects, the client needs to determine a suitable set of contractual definitions that achieve the desired objectives without adversely affecting other aspects of the project. For example, a client concerned with timely completion may incentivise this aspect of the project only to find that the cost has grown substantially and the quality is below expectations, unless both of these areas are also addressed by the incentive scheme.

In order to avoid "reinventing the wheel", clients may want to consider utilising experience from the oil, gas and water utilities to develop contract models to achieve the desirable outcomes.

> The water industry has experimented with new contract models, (IChemE "*Green book*") with recent significant successes, as described in Case studies 6, 7 and 12 (Cleanwater and wastewater treatment plants). These case studies present useful sources of contract information that can be drawn on by parties wishing to adopt an incentivised approach.

Indeed, where projects have been successfully completed using an incentivised approach, there is a real opportunity for each party to gain additional benefits by continuing the working relationship that has been established. This would be particularly appropriate where a whole series of similar works are to be constructed at different locations.

> Case studies 7 and 8 (Water ring main and building refurbishment) identified several projects within a contract where the contractors were self-motivated to achieve the client's objectives since repeat business would likely ensue. Here the incentive for the contractors was the opportunity of more work.

The above are examples of time and cost benefits and the issue of quality has been pre-supposed. During this study, the respondents have generally commented that the achievement of quality, safety and environmental objectives have been considered to be part of the contractor's core management task. The process of incentivisation has to consider the complex interaction of time and cost targets to be met while maintaining these objectives.

Quality, safety and the environment are fundamental areas of concern to the construction industry and they are suitable for consideration under the heading of incentivisation in parallel with other issues. Care must be taken to ensure that such incentivisation does not compromise the responsibilities of the parties to the contract.

> In Case study 2 (Tunnel refurbishment), the client considered that safety was of such importance due to the difficult working environment and the proximity to water that they included the achievement of a zero incident safety record as a target. This was achieved over the period of the contract. The contractor recognised the importance of this target in promoting his organisation with an excellent safety record to future clients.

CIRIA Special Publication 132, *Quality management in construction* (1996), noted that the majority of respondents to its survey "*...cited construction quality as being more important than either construction time or costs...*". Other respondents commented that "*...the subject of quality is becoming increasingly important in the construction process.*"

In recent years, particularly within the water and highways sectors, there have been moves to introduce other forms of payment and risk allocation. On complex projects, the move has been towards agreeing fixed lump sum contracts that transfer most of the risk to the contractors who are also expected to execute projects within tight timescales. Such allocation of risks does not sit well with promoting an alignment of business and project objectives that will result in a "gain-share" situation for both parties.

On straightforward projects, such an approach may deliver acceptable results in the short-term, but it may only serve to focus the contractor's attention on costs, with the likely increase in claims. This would result in clients and their advisors investing heavily in contract administration systems to counter the claims with the potential for there to be an increase in overall construction costs. Incentivisation is likely to reverse this trend, particularly with the growth of DBFO (design, build, finance and operate) contracts under the Government's Public/Private Partnerships Programme where the focus is on optimum risk management as opposed to maximising risk transfer to the private sector.

The majority of the projects reviewed focused on the potentially conflicting areas of cost, time, quality and safety, with a view to achieving a higher chance of a successful outcome to all parties. The results demonstrated both the potential benefits of incentivisation and also that more effort is required to highlight the issues in the tendering stages, so that all parties can properly address them within their respective contract roles.

Summary: the purpose and potential benefits of using an incentivised contract include:

- better performance will either beat the completion date; construction price; quality requirements; improve safety and environmental requirements

- providing a means of aligning clients', consultants' and contractors' objectives to assist in creating a better working relationship

- incorporating a structured risk management/value engineering/ management process as an integral element of any (sophisticated) incentivisation programme – (provided that these programmes engage the input of all parties)

- enabling the gain/share situation to be achieved

- developing focused contract documentation and systems that concentrate on delivering performance rather than protracted administrative procedures

- offering a greater chance of achieving expected or improved financial benefits to both parties.

---

2 HOW INCENTIVE SCHEMES WORK

Incentive schemes can be formulated in a variety of different ways and are traditionally based on a combination of time, cost, and quality. The case studies overwhelmingly demonstrated that time and cost were the main elements leading to benefits for the client and rewards for the contractor for good performance. In addition, any form of incentive scheme should not compromise safety and should allow for costs attributable to safety, such as training. The study suggests that there is some commonality in the environment required, in the processes, procedures and controls adopted and in overcoming the obstacles that are likely to arise. The way in which an individual scheme will work depends on a various factors, principally:

- the initial motivation for the scheme

- the way in which it is implemented

- the industrial sector and type of work involved.

However, there is a common thread: the need for the clients', consultants' and the contractors' objectives to be compatible and remain compatible. It should be noted that the study found that not all of these objectives might be completely compatible given that the businesses of the parties are likely to be quite different. But, if in striving to achieve their own objectives, each party is actively assisting the other to similar success, the incentive scheme has a good chance of working well.

> Case study 12 (Wastewater treatment works) showed that an alternative design proposed by the contractor offered significant cost savings to the client. The use of a construction technique, familiar to the contractor and with the design and design risk accepted by the client, contributed to the better understanding of the risks involved. This responsive approach demonstrated keenness by the contractor to better align the project objectives of all parties. The contract was incentivised at the outset to increase the certainty of meeting time and cost targets. The approach used on this project has encouraged the client to consider other projects in a similar manner.

> Case study 14 (Light processing facility), demonstrated that the client had selected the contractor on its track record of completing new high-tech installations on a fast track basis and was prepared to pay large incentive payments to achieve completion before the due date. This emphasises the fact that the selection of the right contractor is of equal importance to the type of incentive scheme adopted. In this case, it could be argued that the client merely paid for the contractor to resolve any problems, as the principal reward to the client was a contract that was completed as quickly as possible since there was no profit sharing element to the incentive scheme.

The study also highlighted that construction projects are not merely a collection of inanimate tasks; they also include a mixture of people issues and procedural issues that must also be resolved in order to ensure that each project meets the client's requirements and supports the contractor's activities.

To this end, incentive schemes need:

- to be well documented, communicated, understood and agreed by all the parties to the contract – clients, and their advisors, should not resent contractors receiving large bonus payments for improved performance. The scheme should attack costs and not margins. (Indeed there is a case for client site staff and their consultants being included in the incentive scheme, so they benefit from delivery as well as the contractor.) Conversely, contractors should not resent clients making large savings against target costs
- to include realistic and achievable targets which are agreed by all parties – money cannot deliver the impossible
- a supportive environment producing:
  - trust or a clear contractual commitment
  - objectives that are mutually compatible
- to consider the introduction of "drop-dead performance" incentives (no achievement, no money) regardless of what factors may have arisen to prevent achievement.

Two case studies highlighted different ends of this spectrum.

> Case study 14 (Light processing facility) demonstrated that the incentivised payment scheme was aligned with the many key phases of the construction programme. Non-achievement in one phase did not mean that subsequent phases, if achieved on time, would were not attract the incentive payment. This encouraged the contractor to continue to innovate throughout the construction period.

Section 6.2 discusses "hard" and "soft" issues. The potential benefits of the latter may not always be obvious and could be outside any contract. It may be necessary to communicate these benefits to those lower in the management organisation. The potential benefits could be difficult to justify commercially.

> It must be recognised that simply throwing money at a problem will not necessarily solve it. Case study 11 (New power station) highlighted the fact that the civil engineering contractor could not improve on the time delays caused as a result of contamination from a previously demolished structure. This had not been revealed during site investigation. It is in client's interests to use additional measures that incentivise the contractor to undertake activities in pursuance of a successful project completion. Even with the offer of significant cash incentives in this case, there was insufficient time to re-programme the works to meet the deadline date.

Summary: how incentive schemes work

- Alignment of the objectives of all the parties involved, plus the commitment to maintain the integrity of the scheme, all contribute to the successful outcome.
- The mixture of people and procedural issues must be understood to ensure that targets are both realistic and achievable and that risks are properly assigned and managed.
- The processes, procedures and controls must be well documented, communicated and understood.
- Sharing in success should be a common objective.

The next part of the guide will assist readers in understanding the roles and responsibilities of the parties to incentive schemes. It also describes how to overcome some of the obstacles by identifying the fundamental requirements leading to successful application – the "what" and the "why" that are likely to apply.

3 CREATING THE RIGHT ENVIRONMENT

3.1 SETTING AND AGREEING OBJECTIVES AND TARGETS

Chapter 1 demonstrated that the purpose of an incentive scheme is to achieve better performance. The research has indicated that if better performance was to be achieved then there should be clear understanding between the parties about the underlying objectives. This suggests that in general the business objectives of both parties should be compatible with the underlying objectives of the project. It should be noted that apart from issues raised in Case study 11 (Power station), no other evidence was uncovered during the course of the study that suggested there were any cases where either the client's or contractor's overriding objectives prevented incentive schemes from operating effectively. The partnering concept was present in many of the examples considered during the research, but not in all cases. This does not mean that partnering is a necessary element within an incentive scheme; it merely suggests that the parties to the contract may not be considering themselves as partners.

The word partnering is often confused with partnership. The former is a declared intent, orally or with a written side agreement, to work in harmony towards common objectives. It does not override the contractual obligations of the two parties. A partnership is a legal agreement to operate as a single entity for the purposes of the project that states how each party is to be treated, gives rights to each party and, where issues need to be resolved, provides a legal means to do so.

Hence, the absence of any stated partnering arrangements does not rule out the validity of the aim of having complementary if not compatible objectives.

The flow chart represented in Figure 3.1 (Creating an incentive scheme) describes how an incentive scheme may be initiated in a variety of ways. The client or the contractor could start this, either before or during a project. This means that the setting and agreeing of joint objectives can develop in a variety ways. The objectives, however, must be clear, achievable and measurable, otherwise the result will be unpredictable (see also Table 1.1).

Whilst no evidence was uncovered during the course of the study of people abusing incentive schemes to obtain benefits without a commensurate improvement in performance, it is essential for good commercial practice that financial controls are put in place to monitor the performance of clients and contractors. These controls can be executed by the client at one extreme and by contractors at the other extreme, or jointly. Whilst there are likely to be additional administration costs, these should be more than outweighed by the benefits of incentivisation.

The way the objectives are set may be motivated by the client or the contractor, or as a result of common discussion or need. On one hand, the client is likely to have a clearer view of acceptable objectives for a project and is therefore more able to set clear objectives without reference to the contractor. However, there will be a need for the contractor to be involved in assessing the achievability – either by means of a tender negotiation or through a more informal dialogue before the objectives are formalised.

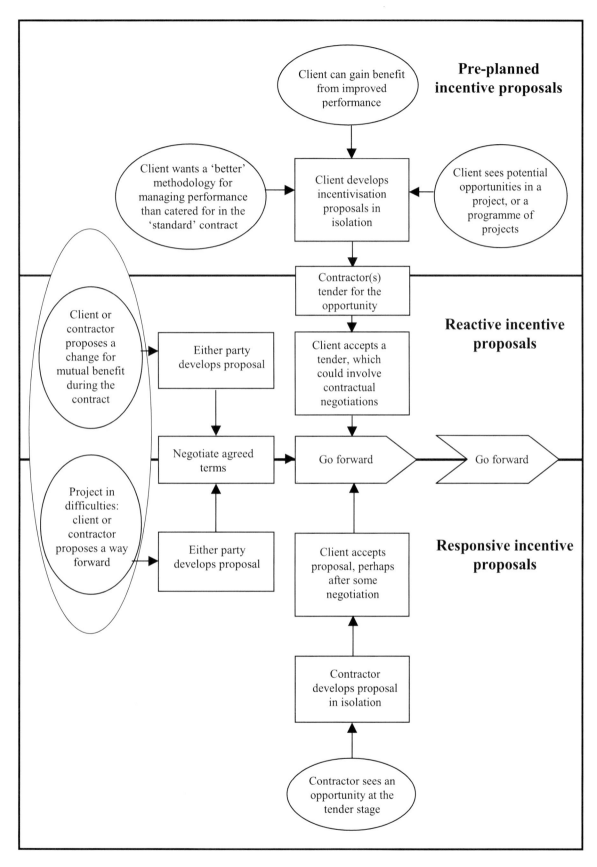

Figure 3.1　　*Creating an incentive scheme*

To ensure that contractors have the fullest understanding, the tender documents must include a full explanation of the client's objectives and values. This will encourage tenders aligned with the client's requirements and ensure that contractors are on board at the earliest opportunity. It should be noted that the introduction of two-stage tendering into the procurement process should clarify who goes forward into Stage II: *this requires a commitment from the client to proceed with a preferred bidder on a single tender action basis, otherwise the remaining contractors may suspect that they are being kept on so that their ideas can be passed on to the preferred bidder.* For public bodies where possible conflict with procurement is a concern the approach should be regarded as a means to improving performance. The Government Procurement Guides 4 and 5 issued by the Office of Government Commerce refer to the development of this philosophy as part of the procurement strategy.

> Case study 12 (Wastewater treatment works) refers to an alternative bid which created an incentive scheme. However they needed assurance that their alternative designs were not "touted" around.

3.1.1 The need for measurement or objectivity in incentive schemes

In the case of new construction projects, the governing contracts can be drafted or amended to encapsulate incentive schemes. However, in the case of existing projects, it is important that any newly introduced incentive schemes are accounted for within the existing contract(s).

> Case study 13 (Road scheme) demonstrated the introduction of a new incentive scheme into an existing contract.

It may be tempting to think in the spirit of co-operation rather than confrontation, that to formalise the incentive arrangements into a contract is unnecessary and contrary to the spirit of trust being encouraged. This is not a route to be encouraged. Properly prepared documents should not constrain either party and informal agreements may still apply.

> There is always an exception to the rule. Case study 14 (Light processing plant) states that at the outset of the building programme, only an informal arrangement for the incentive scheme existed. This was formalised much later on but this approach did not deter the parties from meeting the key completion date.

In essence, incentive schemes are contractual arrangements and of necessity will potentially involve additional, and sometimes early, payments but with commensurate benefits to the client. They should therefore be clearly recorded in documents that both parties agree, set out the scheme in adequate detail. The process of reaching this agreement should be no less rigorous than that followed in agreeing the base contract.

Table 3.1 demonstrates some of the ways different categories of objective may relate to different kinds of project.

Table 3.1   *Applicability of incentive schemes to different types of project*

Objective	Project example
1. Save time against a measurable benefit	An industrial or retail development where the revenue can be brought forward by completing or opening earlier than anticipated. There could be a sliding scale of payment against a specific milestone.
	A highway project where measurable economic benefits flow from early completion. Lane rentals can act as an incentive to contractors for reducing costs by completing highway contracts early and also act as a deterrent for lateness by introducing liquidated damages.
2. Control or reduce costs	Target price contracts where: – benefit on reduced net cost is shared – the return to the contractor(s) can be directly related to their performance.
3. Manage time and quality within a fixed cost envelope	Milestone contracts where payment is tied to completion of an entity. Improved cash flow gives the contractor(s) the incentive to complete early or at least on time. (Quality is formally measured to confirm completion.)
4. Manage time and quality within a cost target	Construction costs may not be the client's biggest issue, eg factory commissioning: getting production on stream without breakdowns is worth more than some overrun in construction cost.
5. Manage time, cost and quality	A turnkey project shifts the appropriate operating and revenue generation risks to the contractor(s) by placing them into a single package for the contractor(s) to deliver. The incentive to complete early, within budget and to a quality, often being the extent of liquidated damages that can be called upon in the contract.

3.2 ATTITUDES AND MOTIVATION

The Latham Report (1994) was written in response, in part, to the recent history within the construction industry for the parties to contracts to become adversarial, leading to confrontation, delays and significant additional costs. It made recommendations for improving relationships and suggested targets for reducing the costs of construction. From the case studies examined, where the client had initiated a different approach, it had been done on the basis that each project needed to be delivered with some certainty of time and cost.

The start of this improvement process and the change in attitudes begins during the tender phase. Case studies in the utilities sector all showed a readiness by clients to stimulate dialogue with the tenderers during the tender phase to produce more robust tenders. In addition wherever possible, it was important that contractors (tenderers) were involved in the design process so that the client was able to draw on their combined construction experience. This allowed the client to identify potential construction problems and then either design them out of the project, or implement measures to mitigate their impact. It also provided an opportunity to discuss the possible introduction of incentive schemes. Case study 17 (Ferry terminals) showed that where complex structures were involved then a design and build (D&B) approach with time and cost targets gave full recognition to the innovation of the contractor and their advisors.

Table 3.2 sets out some key management actions that support the introduction of incentivisation into the tendering process. They are based on experiences from Case studies 1, 2, 7, 10, 16, 19 and 20.

Table 3.2   *Management actions in support of the introduction of incentivisation during the tendering process*

Action	Activity	Reason
1	Define the risks	To ascertain the extent to which risks can be shared under the contract.
2	Agree level of risk transfer with authorising body	To ensure that authorising body is made aware of any potential impact on the contract price.
3	Make a joint presentation to all tenderers	To explain the background to the scheme, with the options rejected and why. To identify known risks and how these could be allocated. To outline other constraints and difficulties and invite suggestions for incentivising the eventual contract.
4	Provide tenderers with as much information as possible, including site investigations, client objectives and constraints	The aim is to ensure that the contractors are fully informed about the scheme.
5	Encourage tenderers to open a dialogue with the project design team on a confidential basis	In order to clarify issues, seek further information, and debate possible incentivisation schemes.
6	Ask tenderers to highlight any anomalies within the tender documentation	To limit any effect on the out-turn price or give a commercial advantage to any tenderer.
7	Encourage alternative designs and invite tenderers to discuss their proposals with the client.	To ascertain acceptability before tenderers commit resources to the development of alternatives. These discussions should be confidential to each tenderer.
8	Lengthen the tender period, if necessary	To ensure that adequate, realistic time is allowed for tendering and subsequent analysis/discussion.
9	Develop a better understanding of the tenderer's view of the construction period and payment mechanism	To reflect the likely outturn costs within the budgetary process to create more meaningful project reporting.
10	Set up a management framework with a flexible approach	To handle reactive issues during the tendering process and during the contract.
11	Maintain effective registers for issuing new or changed documentation	To ensure all tenderers receive prompt updates, taking into account the protection of commercial and confidential considerations.

In general the case studies suggested that tenderers will respond positively to some or all of the above actions, seeing the suggested approach as an opportunity to reduce risk and uncertainty and to submit a tender at a realistic price.

However, there has been some criticism by auditors of incentivised contracts. In the main, this has been due to poorly set up schemes that did not provide the transparency required to audit the process. The lack of a clear audit trail makes it difficult to justify incentive schemes that have been introduced on existing contracts to deal with a crisis. Early involvement of the auditors will smooth the process, as the basis of understanding is formally documented in the shape of a proposal with a response. It is imperative that the decision making process is clearly recorded. Incentive schemes that are set up during the pre-contract stage are likely to be less problematic since the procurement process should ensure a "level playing field" for all tenderers and there is an audit trail inherent in the process.

> Case study 1 (Road bypass scheme) demonstrated an innovative approach to procurement by a public body that resulted in a successful scheme. With direct involvement of corporate functions within the authority and the district auditor, the foundations were laid for ease of understanding of the client's objectives and the alignment of the commercial considerations of the contractor.

The above list of management actions is explained in a more formal manner in a recent working paper of the European Institute of Advanced Project and Contract Management (Epci) (1995). It contains sections on contracting strategies, 1st, 2nd, 3rd and 4th order incentive arrangements. The paper discusses the motivational aspects of compensation (ie payment terms) and comments in detail on contract specific risk premiums and the optimum risk allocation.

1st order incentives must be aimed toward conditions which can be under the contractor's control, eg productivity.

2nd order incentives introduce time as a variable that involves a trade-off between time and cost with a view to reduction of the project's total investment cost without having to obtain a reduction in individual contract time.

3rd order incentives are aimed at suppliers of equipment to ensure an optimal trade off between capital cost, operation and maintenance costs and the unavailability of equipment.

4th order incentives address the trade-off between production capacity and investment costs. In essence this seeks to take account of the whole-life cost of projects.

The paper concluded that research has suggested that contractors can be motivated through profit and pride.

3.3 CONTRACT STRATEGIES

This section is written from the perspective of highly regulated industries, eg oil and gas, since they have advanced the practice of incentivisation as a result of the financial pressures arising from the critical delivery of high output/earning products.

The Epci study (1995) identified two contracting strategies: conventional and alliance. The former included design/build, EPC (engineering, procure, construct), general contractor, multiple prime contractors, project management contractor and a blend of these.

Alliances include supplier alliance, project alliance (see Section 6.1.4) and a blend of the two. The section on 1st order incentives gives a thorough theoretical analysis of contracts, starting with the limitations of lump sum contracts. The authors state the following:

> "a contract target based on lowest bid may for other reasons be less than optimum. For example, in selecting an engineering contractor, the client should be more interested in the life cycle costs associated with construction and operation of what is to be engineered than in the costs for the engineering services per se. Consequently, whole life costs (WLC) of the project are a more adequate target than the lowest bid.
>
> Conventional cost or price-based contract forms are not applicable to contracts with a WLC target for the simple reason that engineering and construction contractors are normally not involved in the operation of the project facilities they design or build. They are neither accountable for operation costs, nor do they have an influence on cost occurring after production start-up. Another difficulty is that the WLC is first known after production termination, which for some projects could be 30 years or more after completing the construction contract.
>
> To overcome these difficulties, the assessment of contractors' performance must be based on anticipated rather than proven results. What we are looking for is how the work of various contractors impacts on our current WLC estimate of the project. At the same time of the project's production termination the actual WLC equals the current WLC estimate.
>
> This way of thinking leads us towards a new concept for controlling costs. Instead of focusing on how cost estimates are turned into expended costs – ie monitoring the consumption of costs – the focus now lies on how the estimated WLC can be continuously attacked in order to be minimised."

The perspective is therefore aimed at the WLC of projects rather than the direct contract costs. Long-term design liabilities and warrantees require that contractors, and clients, need to understand the basis for WLC estimates. This will help to improve designs and long-term performance. The guidance given in CIRIA Special Publication 129 (1996) is relevant here.

Whole life cycle costs are difficult to evaluate as they introduce a number of variables. One professional institution proposes to develop models that will enable WLC assessments to be made with greater certainty. The WLC Forum also has such issues in its objectives.

Following the initial discussion, the authors conclude:

> *"complex contracts with a significant operational elements, eg wastewater treatment plants, should have incentive schemes homing in on how the work influences the project WLC. The anticipated effect of completed contract work could be estimated, for example, on a monthly basis and the result over and above a standard performance remunerated. Financial modelling techniques make this comparison much easier to handle. Through this concept, clients and contractors goals are being aligned. Where the contractor is focussed on improving performance at less overall cost to the client, the lower the anticipated WLC of the project. The achieved reduction in the estimated WLC can be partly deployed to the contractor as a bonus for superior performance. Consequently, we have created a gain-share concept."*

Summary: creating the right environment

- The setting up of an incentive scheme can be initiated in a variety of ways. This is shown in the flow chart indicating that either party can be pro-active.
- For good commercial practice, financial controls should be put in place to monitor performance, both client and contractor. Either or both can execute the monitoring process.
- The need for clear tender documentation is paramount defining the client's objectives and values.
- Properly prepared documents should not constrain either party and informal agreements may still apply.
- Incentive schemes that are set up during the pre-contract stage are likely to be less problematic since the procurement process should ensure a "level playing field" for all tenderers and there is an audit trail inherent in the process.
- The focus should be on whole-life costs as a more adequate target than the lowest bid during the tendering process.
- When the achievement of lower WLC occurs due to improved performance by the contractor then the gain/share concept becomes a reality.

---

4 PROCESSES, PROCEDURES AND CONTROL AREAS

4.1 PROCESSES

The processes involved in considering the use of an incentive scheme should allow the client to both assess the benefits to be derived and to evaluate the risks prior to the identification of the need for incentives to be applied to a part of the project or all of it. Hence this risk assessment would set the size of the incentive scheme to be applied and suggest a gain-share formula that ensures the philosophy is maintained. The evaluation of the cost benefit, giving rise to the incentive scheme, should take into account the whole life costing, as set out in the previous section. The client needs to be familiar with the primary risks and be in a position to facilitate the contractor overcoming these risks.

When the client initiates such risk assessment, then the scheme may be specified by means of a performance matrix built into the payment mechanism.

A detailed risk review is an essential precursor to any incentive scheme. It is also essential that the person responsible undertaking the review has the appropriate knowledge and skills.

The flowchart shown in Figure 4.1 depicts the key decision points in the development of a variety of schemes. The key decisions should be built into the overall control system by the identification of the decision-makers within the authorising body and the supporting processes and procedures that allow proposals to be submitted, assessed and approved. There should be clear separation between the roles and responsibilities of the proposer of the scheme and the approving body such that there is a transparency in how the scheme is to be applied and managed. In more general terms the following high-level decision processes need to be put in place by all parties unless there is a joint management scheme to reduce the administrative costs of the scheme.

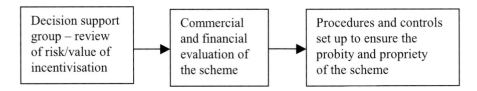

Figure 4.1   *High-level decision processes*

Running in tandem with this decision tree will be the overall procurement process, managed by the client. Care is needed to ensure that they are synchronised to meet the project objectives and approval levels within the client's organisation.

The management information systems supporting the scheme should be set up to complement the financial reporting and project management systems. This should ensure the probity and propriety of the scheme.

Based on the case studies examined, two key process categories emerged:

- the process of determining that an incentive scheme should be applied within a particular contract (or programme).

> Case study 14 (Semi-conductor manufacturing facility) shows that the client wanted a fast-track solution and that substantial bonus payments would be paid if the contractor could complete the factory ahead of schedule.

- the process of determining the element within the contract that should be the subject of the incentive.

> Case study 3 (Escalator maintenance) shows that availability of the escalator is critical to the client's business and that where the contractor is able to properly focus their maintenance resources on key components to maintain availability, an incentive payment will be paid.

4.2 PROCEDURES

There are likely to be four principal categories of procedure involved.

1. The clear identification and agreement of the objectives – this should be set out as a series of management steps so that all parties are responding in a manner that supports the prime objective.

> Case study 8 (Renewal and refurbishment of buildings) shows that the client wished to minimise delay to each project start-up by using a formalised procedure to enable individual project risks to be assessed by the framework contractor. As a consequence, the accuracy of target costs was improved to achieve the client's objectives of completing as many projects within the financial year within budgetary constraints.

2. The development of the documentation in conjunction with an alliance or partnering approach as the words are likely to be capable of misinterpretation or exploitation.

> Case study 7 (Water ring main improvements) shows that following an agreement to develop a framework in which all projects would be procured, the documentation was developed to reflect the continuous nature of the overall scheme. Each and every project required the contractor to provide a target price and share in any savings as a result of improvements to the time and cost of the works.

3. Procurement procedures covering the tender and/or negotiation and the evaluation.

> Case study 1 (Road scheme) challenged the traditional procurement route and sought to develop a new type of document to reflect the innovative tendering arrangements and the receipt of prices and timescales upon which to make meaningful comparisons. This was achieved with the assistance of the district auditor.

4. Feedback of experience – both good and bad – to allow improvement of future schemes where the development of partnering arrangements may be seen as of mutual benefit to all parties.

> Case studies 4, 5 and 6 (Water treatment and wastewater treatment plants) reflect the ongoing nature of these types of treatment plant contracts. The open book style shows that where any significant cost savings could be made they are readily identifiable. In addition operational improvements based on energy consumption and treatment efficiency and effectiveness could be reflected in changes of equipment specifications for future schemes.

4.3 CONTROL AREAS

The study suggested that there are three principal areas that should be covered by incentive (contract) control mechanisms. Table 4.1 lists the areas and examples of evidence from case studies to illustrate the control mechanisms. The purpose of the controls is to enable all personnel associated with the scheme to understand the structure of the scheme and to apply the monitoring process in a consistent manner.

Table 4.1 *Control mechanisms that need to be considered when managing an incentive scheme*

Principal controls	Example evidence
Administration of scheme	*Case studies 4, 5 and 6 (Water and wastewater treatment plants) allowed for a prescriptive administration scheme to operate since this was deemed the most appropriate way for the client to manage the contract.*
Measurement and monitoring of performance and achievement of objectives	*Case study 9 (Motorway-widening scheme) set out a detailed methodology to measure and monitor performance, since considerable cost damages could result if the contractor failed to deliver on time.*
Payment or provision of additional reward to the contractor. (This may not be immediately visible to the client if the incentive is closely bound into the contract's other mechanisms, eg a turnkey project.)	*Case study 2 (Tunnel refurbishment) incorporated a modified payment mechanism whereby the client would release further sums if better performance than that in the contract were achieved. The client team had evaluated the benefits in order to justify the payments.*

In the case of public sector incentivisation schemes, there is greater emphasis placed in applying the controls in a systematic manner so that any individual is protected by the financial regulations and policies of the organisation.

In respect of the private sector, there is a clearer financial trail to the profit and loss account that is part of the reporting mechanism to shareholders.

Summary: processes, procedures and controls

- To achieve the gain/share objective, risks need to assessed by the client and translated into the scheme to maintain the philosophy.
- The procedures to set up the scheme should cover identification and agreement of objectives, development of documentation, procurement procedures and feedback of performance/ experience.
- The controls which should be put in place will need to cover the administration of the scheme, the measurement and monitoring of performance and the payment mechanism.

5 BARRIERS TO IMPLEMENTATION

5.1 CULTURAL, CONTRACTUAL AND MANAGEMENT ISSUES

Within the private sector, the use and development of incentivisation schemes is part and parcel of the commercial arrangements to achieve value for money. Tender negotiations to achieve this result are not subject to such detailed scrutiny as public sector contracts and are held as commercially sensitive by the parties until such times as they wish to release any details.

Within the whole domain of the public sector, and Lord Nolan's Report (1997) emphasised the point, there has to be a transparency of any transaction which enables the process to clearly demonstrate fairness and even-handedness. To introduce incentivisation as part of the traditional procurement route within the public sector requires more openness for the process to become wholly acceptable to internal and external audit.

Public/Private Partnerships (PPP), including the Private Finance Initiative (PFI) is part of a new public procurement methodology, developed to transfer substantial construction and operating risks to the private sector for the reward of a concession contract. The concession is based on life cycle costing and the effectiveness of each element of the project, not just on satisfactory performance on the construction contract. The positive incentive for the private sector is to provide the specified services within a well-defined risk matrix at an affordable price. This price will be lower or equal to the "Public Sector Comparator" (PSC) of procurement by traditional methods. One of the key factors in the use of this procurement method is that extensive discussions on the risks involved in any PPP project create the sense of openness hitherto unknown in traditional procurement. In addition, the accounting regulations associated with these transactions, demands the transfer of substantial risk to the private sector to make the deal acceptable to Government and hence the need for extensive negotiations on risk transfer.

> Case study 1 (Road scheme) demonstrated that the assessment and management of risk in areas of great uncertainty can be managed effectively, but good working arrangements and commitment prior to and during the project are needed.

What needs to be considered as part of the culture change is the development of public procurement methods (such as extended arm contracting and framework agreements) to encourage price competition within a secure commercial environment as part of an overall commitment to the delivery of *best value* services.

Over the past 15 years or so, achieving value for money from public sector contracts was interpreted as seeking the lowest cost tender for the construction of an asset to a pre-determined design and specification.

This can lead to several problems:

- the development of a "claims culture" by contractors to find weaknesses in the contract terms and interpret issues in a strict manner in accordance with those terms. Whilst this is commercially correct, it is being done to recover margins which were reduced or removed as part of the tendering process
- quality being compromised because of a drive to reduce cost and increase (or generate) profit by the contractor
- delays occurring
- minimal investment in R&D by industry as a whole
- escalation of whole life costs as maintenance – free or minimal maintenance options were not sought in favour of cutting construction costs to a minimum.

As noted above, the public sector has, for reasons of public accountability, been reluctant to move away from the tried and tested traditional methods for procuring construction. There are fundamental issues with regard to public accountability that require clear audit trails and management procedures in respect of public money that may be perceived by some as adding to construction management costs. There is no reason why incentive contracts used in the public sector should suffer from such disadvantages if projects that are wholly in the private sector do not. All well-run projects should have clearly defined objectives with visible performance measures.

The private sector has been less constrained by the need to accept of the lowest tender and has appreciated the fact that value may not, and often does not, equate with lowest price.

At the same time greater flexibility in procurement strategy and selection of contractors allows the benefit of value management and risk management to be obtained. These changes have been led by the petro-chemical industries. In recent years commercial property developers have followed and more recently major new privatised companies in the utilities and transportation sections are adopting these approaches.

5.2 INVOLVEMENT OF THE PARTIES, OR IMPOSITION BY THE CLIENT

Changes in procurement strategy have significant impact on the organisational structures of projects, and ultimately on the structure of contracting organisations themselves. This can extend to the outsourcing of services that were traditionally carried out in-house. A good example is the transformation of utilities companies which previously carried out their own project management, design and procurement in-house and have now either become reliant on external consultants to provide these services, or have become reliant on contractor led turnkey project teams. The utility companies have effectively become retailers rather than manufacturers.

Design led construction describes the traditional approach to the procurement of construction projects where the designer/design team are employed by the client. The contractor prices a complete design. Contractors may have to pre-qualify to ensure that only experienced, competent and financially sound companies are short-listed for the final tender stage.

For reasons of public accountability, this arrangement is almost universal in the public sector and it is also widespread in the private sector. It is now being recognised that there are several shortcomings in this arrangement. The Latham report (1994) has identified this shortcoming as a procurement issue which needs to be addressed by clients and the construction industry.

This is not an exclusively British problem since the Epci report (1995) made the following comments:

> "the traditional approach with the lowest bidder winning the contract or order is likely to produce variation orders and claims for late deliveries and delays resulting in cost overruns. There is no alignment of cost drivers and business objectives, and an adversarial relationship is frequently developed. Bluntly speaking, suppliers are considered as 'spot sellers', not as potential partners. In this way both parties deploy a significant level of effort in 'fighting' each other. Suppliers are looking for opportunities to produce claims; clients spend much of their time preparing proactive defence strategies. In some cases the parties come to a cease fire, but frequently they carry on fighting until the disputed issue finally is resolved through litigation."

Traditionally for complex projects, contractors seek to maintain market share through a mixture of commercial realism and professional pride. When this is coupled with the threat of liquidated and ascertained damages claims (LADs) for late completion, submitted by clients, then this is generally considered sufficient to incentivise those involved. However for reasons which have been raised before, these have not necessarily combined to provide the service product of value which clients require. If adversarial conditions develop then the threat of LADS could drive contractors towards extension of time claims.

More recently, insofar as the detailed planning of projects is concerned, many clients no longer have any involvement in the planning of projects apart from setting the overall project periods with start, finish and intermediate milestone dates, ie strategic level planning of projects and integration with overall business and capital investment programmes.

CIRIA Report 85 (1985), under the heading "Programme and method statement", stated:

> "in a target or cost-reimbursable contract the employer may wish to exercise greater control than would be required for either a lump sum or admeasurement contract."

Hand in hand with changes in procurement strategy has been an increase in the adoption of formal quality management systems, usually BS 5750 or an equivalent and their successors, eg the IS0 9000 standards. In theory this has shifted the emphasis from the client and its consultants to check quality, to the contractors and suppliers to first obtain and then retain their quality certification. It is important for the client to clearly define its quality requirements in the context of its business objectives and to audit these. This quality issue has been addressed by the adoption of performance related clauses in contracts where there is a substantial, or total, responsibility on the contractor for design. This has been prevalent in the process industries for many years where the output of plants have been delivered in terms of quantities and composition of outputs. Sometimes bonuses are paid for achieving better than planned performance, eg where innovation in the design process reduces energy and manpower costs, and improved performance is achieved at minimum cost.

5.3 PROCUREMENT METHODS

Traditional procurement routes in the public sector may appear to be tortuous and leave little to the imagination in respect of expediting best value. The development of framework agreements is seen as a means of expediting projects within a traditional commercial framework, but this does not develop incentivisation within the form of contract. What has developed since the Latham Report is a readiness to discuss a more open style of procurement against the backdrop of public accountability.

The recent challenge to traditional procurement strategies within the public sector by the PFI has shown that the constraint on openness can be lifted to encourage the use of incentives within the negotiation process and their inclusion in the contract. Whilst this is an entirely different procurement methodology it has combined the issues of risk and reward within the negotiations and developed more openness in aligning the client and contractor objectives.

In addition, the Government, in its 1998 White Paper on local government services, has set out its requirements as to how Local Authorities must achieve best value services. However, it has not detailed the processes, procedures and controls required.

Within the public sector, during all stages in the development of these processes, procedures and controls, the client's legal and financial sections need to be involved, as well as the district auditor or National Audit Office. This provides the management framework to support the development of "best value".

There are many forms of contract that are designed to meet the needs of particular industry sectors:

- IChemE (Green) with added target mechanisms
- NEC/ECC Options C and D
- NEC/PGC Option D.

They are used because the parties are comfortable with their applicability and overall results. There may be some reluctance among clients to stray outside these traditional boundaries. The development and use of incentives therefore could remain as a sporadic initiative used by those who have taken the time and effort to develop a higher level of accountability within their area of responsibility.

5.4 OVERCOMING THE BARRIERS

The procedures and controls used to systematically manage the use of incentives have had to be developed by each client due to the lack of guidance on their application. The research has shown that they are inexorably linked within the public sector to the audit trail and compliance with the financial regulations. But Procurement Guidances issued by HM Treasury in March 2000, recognise the need for change and have translated the recommendations made in the Efficiency Unit Report on Construction Procurement by Government into practical proposals for implementation. They address the recommendations made in the 1998 Construction Taskforce report, *Rethinking construction*.

Within the private sector, the use of incentives is linked with key commercial objectives and most, if not all project accounting systems directly report these facts.

Given the Government's declared intention (White Paper, 1998) to achieve "best value" across the whole spectrum of public sector procurement, now is the time to promote this awareness by publishing the series of procurement guidances, supported by changes to the Government's financial regulations.

Summary: barriers to implementation

- Within the public sector there is a need for transparency of any transaction which enables the procurement process to clearly demonstrate fairness and even-handedness.
- Cultural change to traditional procurement needs to occur if "best value" is to be achieved.
- Public Private Partnerships have shown a new style of procurement methodology which incentivises the private sector to provide the specified services within a well-defined risk matrix at an affordable price.
- The privatised companies in the utilities and transportation sectors are adopting more flexible procurement strategies.
- These organisations have had to change their structures to manage the new approach.
- Wholesale changes to the procurement method will need corresponding changes to the existing processes, procedures and controls.

Chapter 6 of this publication develops a style of approach for individual clients to use.

6 IMPLEMENTING INCENTIVE SCHEMES

If a contract is be incentivised, then the implementation of an appropriate scheme requires due consideration of all the parties involved to create the gain/share situation. As indicated in Chapter 4, partnering is not necessarily an element within an incentive scheme. The research has shown that it is a management process which is used and where appropriate it is formalised by the parties concerned. The philosophy behind this process is sound and it will require a high level of trust and commitment by all parties, together with appropriate documentation, in order to deliver a successful outcome.

This part of the publication is designed to enable the parties to review the key elements and make joint decisions on which scheme is best for them.

6.1 WHEN INCENTIVES SHOULD BE CONSIDERED

6.1.1 Alignment of client and contractor objectives

As discussed in Section 1.1, it is unlikely that all the objectives of the parties will coincide, but it is necessary for the objectives to be compatible. If this cannot be achieved the reasoning behind the adoption of the incentive scheme should be carefully reviewed to check whether it is likely to be successful.

6.1.2 Timing and structuring the incentive mechanism

The timing of when an incentive scheme arises will depend on which of the parties is initiating it and why. The earlier that the parties can both be aware, at least in outline, of the proposals the better. Based on the incentive categories shown in Table 1.1, the ideal timing is therefore likely to be as shown in Figure 6.1.

6.1.3 Influence of the contract strategy

The integration of incentive schemes into contracts is not a simple matter of amending the clauses related to payments and liquidated damages. There are fundamental issues relating to the allocation of risk, the combination of incentives, planning, cost management, auditing and monitoring of effective performance to be considered.

> Case study 14 (Semi-conductor manufacturing plant) proved the exception to this rule, whereby the prime objective to build the plant as quickly as possible overruled the preparation of tender documents and the normal tendering process.

For these reasons the forms of contract based on the traditional design led procurement strategies may not be suited to incentive schemes. However the study found two examples of these contracts that included a supplementary agreement (Case studies 18 and 20) and which enjoyed some success.

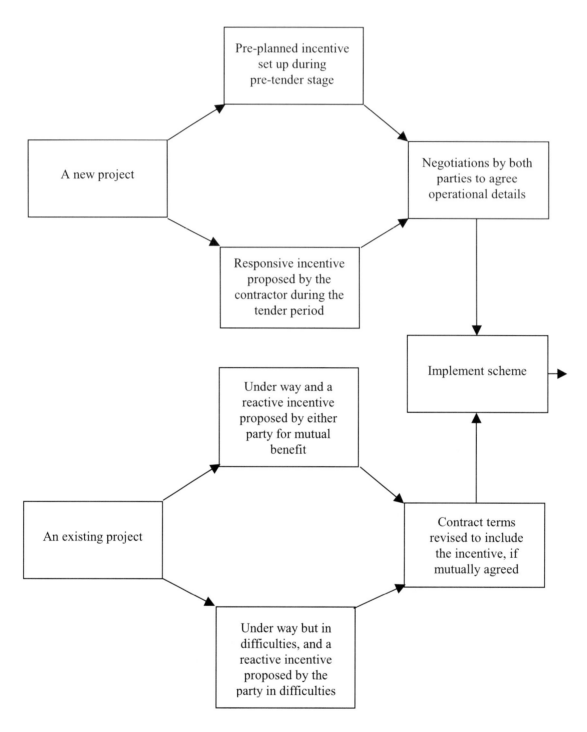

Figure 6.1   *Flow diagram to show timing of types of incentive schemes*

In addition, where there is the use of a "gentleman's agreement" to support the introduction of a scheme then the parties should be aware of potential difficulties that may arise because the fundamental allocation of risk remains as in the base contract.

An article on partnering (Dorter, 1997) has described this problem (not necessarily with incentivisation) as one of :

> "...*failure adequately to recognise whether it is the parties relationships and conduct, as distinct from <u>contract</u> which has been 'partnered".*

The article goes on:

> "...fundamental and finally fatal is the failure to relate the 'philosophy', 'mind set' and 'process' of partnering to the benefits rights and liabilities of the parties under the contract."

In relation to "gentlemen's agreements" it says:

> "such 'moral contracts' and gentlemen's agreements raise the spectre of those Clayton's contracts, whereby each party hoped that the arrangements were a contract binding on the other party but not upon himself."

This reinforces the need identified in the CII study (1986) for incentive programs to be:

> "...perceived as valid measures, flexible enough to meet changing conditions and easily administered."

There is evidence therefore that bespoke contracts including incentive schemes are required, to derive consistently the full benefits.

However, adapting, rewriting, cutting and pasting standard forms with bespoke clauses is not a good route to take. The article referred to above comments:

> "...the plain fact is that simply stapling together the historically relevant documents and correspondence, produces a confusing contradictory collation of paper, likely to lead to disputes."

It follows therefore that appropriately drafted contracts using model forms of contracts such as those mentioned in Section 5.3, are the most suitable routes to forming incentive contracts.

From the case studies examined, the model form most frequently used was the IChemE "Green book" form for reimbursable contracts, which was heavily amended to become target cost contracts. This has the advantage of a significant amount of use by clients, particularly within the water and wastewater treatment sectors. From the general comments of users it was apparent that each incentive scheme required the drafting of specific terms to properly reflect individual requirements.

Within the public sector, changes in contract strategies, through the use of different procurement methods, are likely to be reflected in the greater use of incentivisation. This is simply because improved performance within tight financial constraints remains a central issue. It will be the tried-and-tested incentive schemes which will enable this to happen.

6.1.4 Relationship of "one-off" schemes to framework agreements

From the case studies examined closer integration of the client's and contractor's objectives can be achieved by full partnering types of agreements. These fall into several categories.

The Epci report (1995) defined two arrangements (see Section 3.3):

> The "Supplier alliance" – defined as "*a structure where two or more contractors enter into an ad-hoc alliance contract on a particular project. In the alliance contract the suppliers take the role of an integrated organisation (agent) while the client acts as the principal.*"
>
> The "Project alliance" – defined as "*a structure where the client is included within the alliance group*".

In this case the client is part of an integrated team with common goals and obligations. As with supplier alliances, benefits and risks are being shared according to agreed sharing mechanisms. The overall contract still exists but the way of working is different to create a partnering approach.

The approach being adopted within some Public/Private Partnership (PPP) projects is much more a project alliance than a supplier alliance where the integrated use of current assets and services with new facilities forms part of the overall service provision.

These alliance structures may, as with other forms of organisational structures, be combined.

Other forms of relationships that seek to encourage price competition within a secure commercial environment and demonstrate value for money are through framework agreements and extended arm contracting. These are part of a recent culture change that complies with the EC procurement directives, to establish a series of providers capable of carrying out projects within defined limits and against agreed pricing schedules. Each project is handled as an individual job but it is carried out under the framework contractual agreements. Typically the Ministry of Defence, Rail Passenger Transport Companies and the Utilities use this approach extensively due to the high number of projects to be completed within an annual expenditure programme.

There are greater opportunities to benefit from incentive schemes when applied to rolling programmes of work on the basis of continuous involvement. However, it is vital that the parties concerned do not become complacent.

6.2 WORKING PRACTICES

6.2.1 Contract management and operational issues

From the evidence given in the case studies, it is clear that the full benefits from the successful use of incentivisation schemes are obtained when there is a clear means of aligning the objectives of all parties to the contract. In addition, gaining support from the highest levels within the parties concerned is a prerequisite to ensuring that both management and operational issues are handled in such a way as to secure the desired outcome.

It is essential that any incentive scheme is understood by those who have to operate it. When it becomes necessary to address issues there is advantage in attempting to resolve these initially at the lowest level. In allowing issues to rise up within the organisation, the opportunity for early resolution may be lost.

There was evidence that the development of contract documentation with potential suppliers and contractors before tendering contributed to the principle of "no surprises". This approach gave rise to greater openness and the development of trust.

The importance of communication at all levels within and between the parties cannot be overemphasised and this may require appropriate training to acquire the necessary skills. The client's ability to communicate with the contractor is a key determinant for success.

To focus on the management and operational issues that need to be considered for all incentivisation schemes, there is a need to separate the hard issues (deliverables) from the soft issues (behavioural). These issues will vary in importance and will require different treatments, as each scheme will be unique, reflecting the particular circumstances of the project. This will necessitate changes in attitudes and changes in the criteria that define success in projects. This applies particularly to the soft issues where the benefits may not be obvious and where they may be difficult to justify commercially. The following conclusions can be drawn from the case studies.

Hard issues (impact on deliverables) – key factors that need to be considered are:

- having clearly defined responsibilities
- the use of open book accounting
- how to align business and contract objectives
- agreeing shared risk and reward remuneration basis (incentive schemes)
- how changes are to be managed in a win-win relationship
- what performance criteria and measurement need to be agreed
- how to strive for continual improvement/mutual drive for efficiency
- the development of long-term relationships.

Soft issues (impact on behaviour) – key factors that need to be considered are:

- the selection and chemistry of core team personnel
- how to effect the culture change to move all personnel towards openness
- how business objectives are to be communicated
- the approach to sharing/participating in strategic planning
- the means to carry out team building
- how to develop one team/one culture
- determining goal definition and ownership
- how to communicate a clear understanding of each party's objectives to all levels
- how to fully share data
- determining best value versus excellence;
- focussing success on asset criteria – not on individual areas
- installing transparent boundaries/no corporate barriers
- getting the job done in the right place rather than your place
- asking questions of each other (be curious)
- listening without prejudice
- having flexibility
- expecting people will do what they said they would do
- attempting to avoid duplication
- minimising paperwork

- getting it right the first time
- believing in what you do
- developing a team spirit.

6.2.2 Additional liabilities to be considered

In the case of "supplier" and "project" type contracts there are several liabilities that have to be considered:

- responsibility for defective design/workmanship
- contribution to knock-on effect of defective design/workmanship
- delay in delivery of design/fabrication/equipment
- consequential losses
- overall cap on liability
- parent company guarantees and/or performance bonds
- third party liabilities.

These liabilities also imply changes in tendering procedures. Good practice in selecting potential contractors is explained in CIRIA's companion report, *Selecting contractors by value* (1998). However, insofar as incentive contracts are concerned, the following points require emphasis:

- the contractor must fully understand the client's objectives and these must be transparent to the tenderers.

- all incentive schemes must be clearly explained or jointly prepared and applied in a responsible manner.

6.2.3 Adopting good working practices

CIRIA Report 85 (1985) emphasised that:

> "...whatever strategy is adopted, the contract documents must be concise, unambiguous and must give a clear picture of the division of responsibility between the parties. Risks should be identified and the document should provide for each statement of how it is to be apportioned and the relevant method of payment and / or target adjustment."

On the subject of assessment of tenders, it sounded the following warning (ibid):

> "before the contract is awarded, it is essential that the employer (or his engineer) scrutinises the contractor's proposals to ensure that all parties have the same view of problem areas and scope of the work, and to identify and overcome any misunderstanding. In several observed contracts scant attention had been paid during the assessment period to the method of working proposed by the contractor and this led to unnecessary conflict and early adjustment of the price or target when changes were required by the employer soon after the award of the contract."

CIRIA's companion reports, *Value management in construction: a client's guide* (Special Publication 129, 1996) and *Control of risk – a guide to the systematic management of risk from construction* (Special Publication 125, 1995), provide good practice guides to these subjects.

Greater benefits of these processes will be derived if the input of contractors (tenderers) can be obtained early in the project life cycle.

Health and safety must always be a paramount consideration in construction that due to its nature is, to an extent, inherently hazardous. Therefore it goes without saying that the incentivisation of cost reduction or early completion must not be allowed to compromise safety in any way. In fact, any form of incentive scheme should not compromise safety. More specifically, any bonus earned could be factored by the safety performance. Multiplying the sum due by the frequency for lost time accidents is a useful way to apply this principle. Ideally the figure heads below 1.0 and in poor examples heads to 1.5 or higher, which automatically reduces the payment due where safety performance is poor.

> Case study 2 (Tunnel refurbishment), demonstrated the greater significance of safety issues due to the proximity of water to the excavations. Focussing the safety performance issue as a key non-financial indicator highlighted the mutual objective and achievement of an exemplary safety record.

6.2.4 Tools and techniques for success

To enable schemes to be implemented as effectively as possible, there are a number of tools and techniques that can be applied to the early phases of a project. The degree of buy in by the parties and the success of the scheme in delivering the works can be reviewed throughout the project and adjusted as needs be or dropped or improved for other phases of the project.

Toolbox 1 (Glossary of terms) describes the terminology that is frequently used in incentive schemes. This is not an exhaustive list.

Toolbox 2 (When should incentives be considered?) sets out an informal checklist approach that raises questions to help identify all known problems and how they should be considered in the context of the scheme. Recent additions to procurement methods as part of the Public Private Partnership initiative have highlighted the opportunities to all parties to consider the longer-term consequences of risk transfer. In the context of the negotiated procurement process, there is considerable activity during the early stages of a project to determine what sort of contractual arrangement will overcome any difficulties between the parties over the contract period.

Toolbox 3 (Applying incentive schemes) lists a series of key questions that need to be addressed as part of setting up the scheme, the operational environment and how the performance is to be measured. In addition, it highlights the feedback loop which will provide the means by which continuous improvement can be achieved. These matters are defined in greater detail in the next section.

6.3 HOW TO IMPLEMENT AN INCENTIVE SCHEME

6.3.1 Setting the scene

As more and more interest is now focussed on the use of incentives within contracts, there is a general acceptance that incentive arrangements should be considered to deliver better value for money. As more experience is obtained in their operational use with more successes rather than failures, then they will become the norm. The fundamental approach by the promoter of the scheme should be to encourage the participants to work together to eliminate:

> "*wasteful activities that do not add value to the project and to identify and implement process improvements, alternative designs, working methods and other activities that result in added value to the successful project*".

> (Procurement Guidance, 2000, circulated by HM Treasury for consultation).

To put the promoter in a positive frame of mind about incentivisation, they should consider the objectives of the scheme as if it were part of the process of setting up a partnering arrangement. This creates an environment to identify the key project objectives and to build a contractual framework for these to be achieved. If the incentives form part of a separate partnering arrangement, then the base contract will need to recognise this arrangement and not be in conflict. The terms and conditions of the base contract will also need to be modified to reflect the new way of working.

6.3.2 Setting the targets

Where, for example, the client is taking the initiative through the pre-planned route, the broad elements of the scheme will be set out on the basis of achieving the key parameters of time, cost and quality. For different types of project, there will be different targets which reflect the importance of the client's objectives. The case studies have demonstrated that where, for example, high value end products are involved, time is of the essence and consequently this target has the incentives applied to it. The cost target in these circumstances remains important but is subservient to the time target.

6.3.3 Establishing the incentives

Incentives should be set up so that the party most closely associated with achieving the significant improvement in performance, is rewarded accordingly. If this is a sub-contractor or sub-consultant then the main contractor or main consultant should not necessarily assume that they would also be similarly rewarded. However with the gain/share culture, the promoter should be mindful that all parties to the contract contribute in some way and there may be a case for a graded incentive scheme to be put in place to recognise the varying contributions.

Throughout this period of incentive setting, there is a trade-off between financial reward and the achievement of the project objectives. Part of the failure of some of the confrontational contracts of the past is that the risk/reward equation was biased towards one party or another and consequently this did not create the right environment for innovative thinking.

There needs to be a measure of flexibility built into the incentivisation process. Where circumstances change on the project which is not attributable to any party, then there should be a mechanism to review the incentives originally set up and make changes as appropriate. Changes can and do occur within the client's organisation which may require a radical rethink of the project. Again these mechanisms need to be clearly identified to re-align the project to the new direction.

6.3.4 Managing the processes

As with all challenges to the conventional way of working in the public and private sectors, the use of incentives in contracts requires a new way of working. It is recommended that the new processes are mapped to clearly identify the many interfaces and responses needed to ensure that the incentive scheme has been properly set up with all the principal parties involved. There are many software tools on the market which have this capability, and the diagrammatic representation of the flow of information and decisions will help all those involved to understand their role and contribution to the overall objective. Several of the organisations represented within the case studies had developed standard documentation and formats to handle incentivised contracts.

6.3.5 Applying the tools and techniques for success

The guide has set out two checklists of activities (Toolboxes 2 and 3) which if followed will enable the promoter to be more confident that he has raised all the appropriate questions prior to entering into an incentivised contract and/or partnering arrangement. The essential ingredients within the incentive scheme will vary form project to project, but within the various sectors of the construction industry, there are similarities in approach, which was evidenced by findings from the case studies.

The short list of dos and don'ts as set out below may provide a quick, ready reference to setting up an incentive scheme.

- Do carry out expert value management and risk management exercises and ensure that all stakeholders are aware of the issues and have contributed to the discussion.
- Do identify the targets which need to be incentivised, setting aside those which will be normal contractual commitments/obligations.
- Don't set incentives for improvements in performance that are of no value to the client.
- Do ensure that all parties to the contract (client, contractor and other consultants) understand the incentive scheme and that any potential "blockers" to its success are identified
- Do identify how the improved performance can be adequately measured and monitored.
- Don't set unrealistic targets, such as "drop-dead dates" which if not achieved completely demotivates the Project Team without giving them the opportunity to subsequently recover the situation at a later stage.
- Do ensure that the right people with the right behaviours are involved with the project; clashes in personalities can cause projects to miss achievable targets.
- Do enable tenderers to prepare innovative schemes and methods to achieve project goals and guarantee their confidentiality.
- Do carry out some preliminary short-listing process to ascertain the most appropriate tenderers for the project.
- Do develop the chemistry between the client and the successful tenderer.

- Don't follow the same pattern each time, but review projects within the broad principles of incentivisation. It is important that selection criteria are reviewed for each project.
- Do feedback positive and negative issues arising from the project. This will create the opportunity to review where mistakes are frequently being made and what corrective action can be taken for future projects.
- Do share experiences within industry sectors.
- Above all do ensure that the principle of gain/share is supported within any incentivisation scheme by robust processes, procedures and mechanisms which can handle changes in circumstances.

6.4 PERFORMANCE AND FEEDBACK

6.4.1 Monitoring scheme performance

In order for incentive contracts to be implemented, the achievement of the client's objectives for which additional payment is due must be measured.

As noted above, CIRIA Report 85 (1985) investigated and commented on the issue. The CII study (1986) also commented that the performance measures used in an incentive programme should be capable of being easily administered. There is evidence that time targets are easier to administer than cost targets. This is only true if the scope of work, the degree of completeness and the time are very clearly defined. Quality targets and defects/breakdowns are specific targets which can be measured and are worthy of consideration within schemes. CIRIA Report 85 (1985) stated that for cost targets:

> For certain types of incentive projects, *"the contractor's accounts must be open to scrutiny by the employer, who will need to establish monitoring and audit procedures to ensure that all entries are properly incurred in the completion of the works defined in the contract. Audit should proceed and be completed concurrently with construction work"*.

This can absorb resources and it would be more beneficial if one party carries out the cost checks/ record keeping (trust) and the accounts periodically audited (nearly trust). There was evidence on water utility projects (Case studies 4, 5 and 6 – Treatment plants) that where profit margins are low, the level of administration is a sufficient financial burden on all parties to cause the questioning of the validity of the approach. The possibility of follow-on work for the contractor was the one of the principal incentives in these cases. In the case studies mentioned the client was willing to protect the contractor's profit margin (assured to be reasonable by the open-book approach) thereby allowed all parties to focus on driving out excess costs ie not the usual client threat of attacking margins.

The administration of time targets is relatively simple. In particular, the parties are not involved in the additional measurement and evaluation normally associated with a cost target: adjustment of a time target would be by any extension of time allowed under the relevant conditions of contract (excepting any regulatory/mandatory targets that have to be met).

Another feature of target cost contracts is the need to adjust the target to maintain the incentive element when changes occur. The rules for change can be complex and not all extras add to the target, eg defective work. Such changes may include altering the scope of works. Other issues that should be considered include payments for risks that the contractor is not responsible for, unforeseen conditions and other rules that are outside or excluded from the contractor's responsibility. CIRIA Report 85 (1985) recommended that a simple method of adjustment is required.

6.4.2 Potential benefits and apparent disadvantages

The potential benefits of using incentive contracts include the:

- closer alignment of clients and contractors objectives
- facilitation of a gain/share situation
- greater understanding of each party's position
- greater co-operation in planning and administering contracts
- need to clarify and understand objectives, budgets and programmes.

The apparent disadvantages include the following:

- quality can be driven down to meet cost and time as these are the two aspects most often associated with incentivisation
- both parties at the pre-tender (client) and tender (contractor) stages are required to apply more effort to complete the tasks
- additional effort may be required to audit the project
- to be effective, multiple incentives are likely to be needed. These require care in determining priorities, designing the incentive controls and understanding the interaction between them
- to be effective, realistic targets must be set at the outset of the contract, hence tender assessments must be thorough and commercial negotiations may be necessary to agree appropriate figures
- some cultural changes may be required (although this maybe considered a benefit)
- to be fully effective, subcontractors and suppliers may need to be involved in the tender and negotiations. This may be difficult to arrange
- a pre-determined incentive scheme can turn out to be liquidated damages in disguise, however a share of the overruns may be a more moderate approach.

It is generally recognised that the additional effort during the early stages of projects is beneficial in achieving the client's objectives. Good faith, trust and flexibility are required. This necessitates changes in traditional attitudes and work patterns. Nevertheless these are likely to be less difficult for incentivised contracts than for full partnering arrangements.

Indeed the formulation and agreement of incentivised contracts may be viewed as a stepping-stone towards full partnering.

6.4.3 Establishing sector model contracts

Since it appears from the research that there is limited cross-over of application of the detail of incentive schemes between the different construction sectors, the most fruitful area for the development of contract models for incentive schemes will be within the same sectors.

This will allow potential users to identify those incentive schemes that have been successful within particular sectors – which will be the models that are most likely to be successful if followed closely.

It does not preclude potential users from looking at any sector and if they see a format or set of circumstances that may be transportable and adaptable to their needs, developing a model from another sector may provide the best solution.

Initial dissemination of such models is likely to be by means of articles in the technical press and by the use of professional advisors to the client.

Care is needed in drafting incentive schemes:

- is a target date moveable by extension of time or is it fixed regardless?
- are target costs moveable by variation orders or are they fixed regardless?

Both should not have a payment to finish on time or to cost, only to be better. Both should be on a sliding scale and specify what happens if they fail.

In due course, it is to hoped that bodies responsible for the development and maintenance of contracts will pick up the models that are proving most successful thereby providing the opportunity for their use within the standard formats, at least by way of guidance notes.

Summary: implementing incentive schemes

- When incentives should be considered:
  - where there is an alignment of client and contractor objectives
  - when the timing is appropriate to enable improvement in performance to provide better value for money
  - how to influence the contract strategy with appropriate changes to the terms and conditions of current contract forms.
  - the extent to which framework agreements cover the incentives needed.
- Working practices:
  - understanding how behaviours at all levels within an organisation can influence the success or otherwise of the scheme
  - consideration of additional liabilities
  - adopting good working practices
- How to implement an incentive scheme: consideration of the dos and don'ts
- Performance and feedback: monitoring scheme performance.

PART 2

INCENTIVE SCHEMES: THE CASE STUDIES

BACKGROUND TO CASE STUDIES

This appendix contains survey information collected during CIRIA's study of experiences of using incentivised contracts in construction. A total of 20 projects were reviewed across a wide spectrum of the construction industry. Key interviews were held with client and contractor representatives to determine their experiences in using an incentivised approach. The information is presented in the form of case studies. Each case study pro forma contains:

- basic information about the project
- the form of contract and description of the incentivisation scheme
- identification of measurable benefits
- experiences in operating the scheme
- experiences of how each scheme was developed and managed
- identification of possible improvements
- summary comments from the research team.

The survey was carried out in 1997–1998 and included clients, consultants and contractors from the following industry sectors; utilities, transportation, civils infrastructure, buildings, heavy process manufacturing and light process manufacturing.

OBJECTIVES OF EACH STUDY

The objective of the research programme was to assess the success or failure of using the incentivised approach within the contract to improve the likelihood of meeting the primary targets of time, cost and quality.

Each study sought to identify, evaluate and draw conclusions from:

- the benefits obtained by clients and contractors in the various sectors who had followed this procurement route
- the modifications to standard contract documents and resources required to set up and operate schemes
- the lessons learned from the experience which could be applied to future contracts.

FIELD RESEARCH

Based on a structured interview questionnaire, the research was carried out by a series of interviews with parties associated with incentivised contracts on 20 projects. Over 50 interviews were held with clients, consultants and contractors retrospectively after the projects were completed.

The projects within each sector were selected to reflect the broad use of incentives.

The interviews were generally conducted in the interviewee's office and the responses to the detailed questions were used to gauge the level of commitment and attitude of the people involved. The interview notes were consolidated into a case study format that would facilitate comparison of events. The extent of the findings obviously refers to the number of projects reviewed, but the collective experience of the Steering Group members gave added weight to the arguments put forward.

The essence of the interview was to seek to identify:

- reasons for using an incentive scheme
- the expected benefits from using an incentivised contract
- the benefits actually achieved.

In some cases, other projects were discussed, to demonstrate that lessons could be learned from these experiences to give added impetus to the use of incentives.

The key findings are presented in sections that broadly follow the order in which information was obtained in the research. The comments offered by the research team generally reflect the tone and enthusiasm of the interviewees.

To assist the reader we have reproduced Table 1.1 below, explaining the three main categories of incentivisation.

Table 1.1 *(reproduced)*   *Incentive categories*

Category of incentive	Stage or type of project	Likely motivation	Way incentive is proposed
Pre-planned	A new project or programme of projects	Client wishes to obtain "better" performance or gain benefit from improved performance	By client before contractors are invited to tender or negotiate for the contract
Responsive	A new project	Contractor trying to win a contract by proposing an added benefit for the client	By contractor during tender
Reactive	A project that is under way	Either the client and/or contractor see a mutual benefit by proposing a change	By either party during the construction, once the opportunity becomes apparent
	A project that is underway but in some difficulties	Either the client and/or contractor trying to save a project that has run into difficulties from technical aspects	By either party during the construction, once difficulties become apparent

CASE STUDY 1

Sector: civils infrastructure	Project type: road scheme – bypass
Project description: a complex scheme consisting of 900 m length of single carriageway road with numerous structures. Previous heavy industrial use of the site had left the whole area contaminated.	

Objective of the incentive scheme: to obtain some certainty with regard to time and financial outturn.

Background information: a similar project on a nearby contaminated site had been contracted on a traditional basis. The relationships between the parties had developed into a confrontation over claims, which had led to litigation. The client was keen to avoid a repetition of these events and set out to improve on standard performance by introducing a contract which encouraged the achievement of common objectives and goals with the aim of meeting the prime objective above.

Incentivisation category: pre-planned	Overall project value: £10.1 million
Overall outcome (success/partial success/failure) Success. Both parties considered the tendering procedure and contractual arrangements had significantly contributed to the successful outcome of the project.	Measurable benefits (time/cost/quality) Expected – 15 month contract. Within budget. Achieved – Completed a few weeks early. Outturn price within client's budget. Contractor made a profit.
Form of contract: ICE 6th edition – standard form with a complete revision of Clause 60 (Payment Mechanism Clause).	

Description of scheme documentation: interim valuations were replaced by a tendered payments profile. This enabled the contractor to determine their own positive cash flow profile without the need to adjust tender rates to generate payment in the early months of the contract. The intention of this initiative was two-fold:

1. To free up site staff time to deal with the valuation of variations rather than undertake monthly interim measurements

2. To achieve tendered Bills of Quantities rates which reflected the actual cost of undertaking the items of work by allowing the contractor to generate a positive cash flow without the need to resort to loading rates for work occurring in the early part of the job.

By separate agreement: none.

Operation of the scheme – administrative resource needed: both parties agreed at a post-tender workshop that there was a need to minimise site correspondence – a system of internal correspondence was introduced based on a standard form and common filing system. The shared site office and reception contributed towards positive working relationships.

Evidence: summary of views expressed in interviews/documents released

Client/consultant: the client's previous contract experience in the use of the traditional approach on an adjacent site had convinced them of the need for a better alternative. A corporate decision was taken to change the normal tendering practice to enhance the information provided and to actively encourage alternative designs. This action was taken to reduce the likelihood of disputes arising on this difficult site.

The innovative tendering procedure was adopted to increase the likelihood of the project being delivered with some certainty as to time and cost. The tendering procedure was amended in several important ways:

1. To stimulate dialogue and provide as much information as possible to enable more accurate tenders to be produced.

2. Alternative designs were encouraged and discussed in confidence.

3. The tender period was lengthened to 3 months to allow more time for discussions.

Contractor: due to UK wide difficulties within the construction sector, the Board had decided that a significant culture change was required within the company to bring about a new management style to survive. This culture change used the "customer focus" as a key objective on all tenders. This project allowed this new style approach to be fully tested and implemented. Firm commitment was from Board level downwards. Previous experience with this client had been on a confrontational and adversarial basis.

How was the incentive scheme developed and handled: as a result of the discussions held during the tendering period the following key factors were built into the contract as incentives to achieve improvements on standard performance:

- construction time estimates were determined by the successful tenderer's time required for the agreed design which was contained within a generous maximum set by the client.

- interim valuations were replaced by a tendered payments profile – positive cash flow encouraged.

- use of value engineering with an agreed procedure for sharing cost savings.

The County Council's Legal and Financial Sections were kept fully informed as was the District Auditor who vetted the winning tenderer's profile in detail before acceptance of the tender.

Management of the incentive scheme

- An initial two-day workshop was set up to foster harmonious relationship throughout contractor, client and consultants organisations. A series of common objectives were agreed, including time cost and quality as well as an impeccable safety record.

- Continuous process of contract review was set up to forecast any likely problem areas and to put forward likely solutions. A minimal site correspondence process was installed using a "partner memo" (an internal memo style rather than formal letters) together with a shared office and reception.

- Daily problems sorted on a shadow partner basis ie paired members from engineer and contractor, empowered to take decisions.

What improvements could be made on future projects of a similar nature

- To most clients the avoidance of time delays and cost overruns remains a prime objective. Whilst the transfer of all risks to the contractor appears to be the intent of many clients, this can lead to serious contractual disputes. This is to be avoided. The identification of all significant risks at an early stage in the tendering process is the first step. This process supports the alignment of client and contractor objectives and sets up the opportunity to use incentivisation as a means of improving performance.

- It is vital that there are frequent reviews of the management arrangements to ensure that the process is working correctly. Using a workshop approach maintains stakeholder involvement and contributes towards positive working relationships.

Any additional information for future reference

- The client considered that the potential risk of time and cost overruns warranted revised contract arrangements. With the risks identified and priced accordingly, there was the opportunity to incentivise better performance through the payment mechanism.

- The management of the contract was structured to give the emphasis to finding the most cost effective solution before financial responsibility was considered. This encouraged meaningful and productive dialogue with all parties.

- The approach followed allowed staff to do the work that they were trained for, ie constructing the works.

Research contractor's comments

- Both parties agreed that the new approach enabled a 100 per cent achievement of the objectives and has set a precedent for future works of this nature where the amount of risks are high.

- Using an innovative tendering process has reduced the likelihood of any significant risks remaining undiscovered. This encouraged the contractor to seek potential cost savings knowing that profitability was safeguarded. Potential cost savings of over £250 000 had been identified within the tender sum.

The lessons to be learnt from this project

1. Where previous projects have failed to deliver the desired objectives using a traditional approach, then the opportunity to change the form of contract and the procurement methodology should be taken provided there is a willingness by all parties to seek and implement radical solutions.

2. The different management style and commitment from both parties to make the process work enabled high risks to be handled and for improved performance to be adequately rewarded.

3. The most telling comment from both parties who had been involved on the adjacent site was that there had to be a better way to procure and manage high risk projects. This provided the impetus to set in train the revised approach.

CASE STUDY 2

Sector: transportation	**Project type:** tunnel refurbishment
Project description: the strengthening of the world's first shield-driven and first significant sub-aqueous tunnel completed in 1843.	

Objective of the incentive scheme: to improve on the target time and cost

Background information: prior to the Grade II* listing of the structure, analysis by the client demonstrated that the tunnel was in urgent need of strengthening in order to extend its working life. The history and uncertain structural integrity of this brick tunnel presented the project team with an unique set of problems to overcome not least amongst them being the task of accommodating the diverse views presented by a number of experts from within the industry.

Incentivisation: reactive – mutual benefit	**Overall project value:** £23 million
Overall outcome (success/partial success/failure) Success. A high quality end product with a value-for-money solution to the combined needs of the client and English Heritage. Achieved through an effective design and construction process.	**Measurable benefits (time/cost/quality)** **Expected:** 50 week programme. Within negotiated price. Tough safety targets. **Achieved:** Completed a week early within budget and with no outstanding commercial issues. Excellent safety record.
Form of contract: ICE Design and Construct with Clause 12 included (unforeseen circumstances).	

Description of scheme documentation: a modified payment mechanism was incorporated using milestones and programmed dates. Following appointment of main contractor with a nominated structural designer and consultant engineer for drainage and flood barrier design a number of scheme options were investigated.

Extensive feasibility and value engineering studies, over a period of one year, were conducted which resulted in the agreement on an "open book" basis of the contract price and programme. Against this target cost, for each day over the agreed completion date, liquidated and ascertained damages (LADs) of £11 900 per day would be sought by the client.

The contractor also proposed that a terminal bonus (equivalent to £11 900 per day) would be paid for early completion up to a maximum of 21 days. This was agreed upon by the client.

By separate agreement: none.

Operation of the scheme – administrative resource needed: the level of resource needed was minimised by the close working relationship between the client, main contractor and nominated structural designer. This partnership was based on the mutual trust and respect of individual team members and on the long-term relationships built up between these organisations over many years on a large number and variety of other projects.

Evidence: summary of views expressed in interviews/documents released

Client/consultant: following an initial tendering exercise with a traditional form of contract it became evident that considerable interest in the heritage aspects of the project could jeopardise the commencement of work. To comply with English Heritage demands and the listed building consent, the client negotiated the contract based on a full risk analysis to a lump sum target price.

The negotiated price had milestone payments and identified a bonus payment for early completion. The one year of site investigations, whilst delaying the commencement date, did provide more time for detailed risk analysis and value engineering. Good staff and working relationships helped considerably in this complex and difficult project. High safety requirements were rigidly applied.

Contractor: a traditional approach without Clause 12 would have placed considerable risk with the contractor. The extended negotiations with English Heritage allowed minor enabling works to proceed. They also gave more time to re-appraise the techniques and finishes to the tunnel lining and provide innovative solutions. With an open book approach to the negotiated target price most of the risk had been identified and assessed. The absence of any large undiscovered risk elements led to a better focus on performance rather than compliance with traditional contract mechanisms.

The opportunity with the change in construction process meant that specialist sub-contractors were engaged to become full members of the integrated project team.

How was the incentive scheme developed and handled: during the extensive discussions with English Heritage prior to the commencement of the contract works, a detailed risk assessment enabled the contractor to propose an alternative design as part of the price negotiations. This negotiation took place under open book arrangement to ensure that the final lump sum price was agreeable to both parties. The bonus payment was set out as a terminal arrangement to provide an incentive for accelerated performance if practicable.

The negotiations were handled at board level by both parties. The commitment to achieving a successful outcome emanated from this point. The scheme was introduced in response to changed circumstances when the project was encountering problems. It was tailored to suit the new circumstances to the mutual benefit of both parties.

Management of the incentive scheme: with the negotiated lump sum fixed, the book was closed and the contractor was set the target to complete the work within this price (subject to agreed variations). Milestones and programme dates were proposed by the contractor and agreed by the client. Their achievement by the contractor allowed payments to be released. This approach encouraged resolution of problems to maintain target performance.

The change in approach meant that the project became essentially a programming and planning activity to meet milestone performance. More flexibility was introduced to facilitate sub-contractor achievement of objectives. The incentivisation moved the focus from evaluating risk to one of monitoring performance.

The achievement of an exemplary safety record was a mutual objective actively promoted by the client and the contractor with the key sub-contractors in support.

Since the necessary waterproofing and trackwork activities had to be carefully phased in with tunnelling operations, the key subcontractors went through a detailed selection process to ensure that they would become full members of the integrated project team.

The robust nature of the working relationships ensured that areas of potential conflict were discussed and resolved before resorting to the contract with all parties working to the same goals of quality, safety, productivity and value-for-money.

What improvements could be made on future projects of a similar nature: the risks associated with tunnelling projects involving old structures need to be adequately assessed prior to the invitation to tender. Where insufficient time is given for any Tenderer to fully identify and evaluate the tunnel lining risks prior to submitting a Tender, then there is greater potential for claims and disputes to arise. This suggests a greater emphasis should be placed on providing detailed risk assessments as part of the tender documentation to provide more meaningful tender pricing.

Any additional information for future reference: where risk has to be priced then it should only be done within the context of a management regime to handle the risk. Performance should be incentivised based on known management of risks.

Research contractor's comments
- By carrying out extensive investigations on the existing tunnel lining prior to negotiating the target price, the significant risks of time and cost overruns were mitigated. This enabled the contractor to price for performance and not risk, with the proposed target price vetted by the client.

- Both parties were then focussed on performance which enabled a 100 per cent achievement of the objectives. Improvement on the completion date was always under consideration by the contractor with a terminal bonus as an incentive. In addition this encouraged the contractor to seek potential cost savings knowing that profitability was safeguarded.

The lessons to be learnt from this project
1. Where significant construction risks are likely to occur, then extensive site investigation, whilst time consuming and costly, contributes to the alignment of client and contractor objectives during the tendering process. The tender can be priced accordingly and performance can be incentivised.

2. The close management style between client, structural designer, main contractor and sub-contractors enabled difficult technical problems to be overcome, to maintain target performance.

3. The involvement of English Heritage as an independent party approving and overseeing changes and additions to any listed structure acted as a catalyst in cementing the working relationship between client and contractor.

CASE STUDY 3

Sector: transportation	**Project type:** escalator heavy maintenance
Project description: the maintenance of escalators is carried out by several contractors each working within a framework agreement with the client. The provision of support services is operated by specific agreements using a performance-based specification.	

Objective of the incentive scheme: to return each escalator to service by its programmed date, within the target cost and within the specified safety and quality parameters.

Background information: with over 300 escalators of different types and varying ages and an operating regime of up to 20 hours per day, seven days per week, the demand for minimum downtime presents a significant care and maintenance challenge. A dispersed contractor's staff with a large amount of plant and equipment is required to be available to carry out heavy overhaul maintenance works. The assets have suffered through the lack of capital investment over many years although the aim is to continue to strive to reduce the high cost of reactive maintenance using a mix of the capital renewal programme and heavy overhauls. Most of the travelling public are well aware of the difficulties faced with the non-availability of escalators at deep-tube stations. Time overruns have a greater significance than cost overruns.

Incentivisation: pre-planned	**Overall project value:** various, annual spend £4–5 million
Overall outcome (success/partial success/failure) Partial success. The client and each contractor have a difference of opinion as to the most effective procurement methodology. There are only a few competent global contractors and they dominate the marketplace. This is recognised at board level by all parties. The lack of capital investment constrains what can be achieved.	**Measurable benefits (time/cost/quality)** **Expected** – to specified programme return date. Within target cost. Within set quality and safety parameters. **Achieved** – generally time overruns but within target cost. In general safety and quality parameters met.
Form of contract: a bespoke contract modelled on a recent successful escalator renewal and ten-year maintenance project. Historically this has developed from the model form issued by the National Association of Lift and Escalator Manufacturers.	

Description of scheme documentation: the client has developed a prime cost definition and target cost formula which seeks to reimburse the contractor for the net cost of the works plus a fixed fee, contained within an overall target cost. There are bonus payments for any improvement on the finally adjusted target cost. The bonus payment is abated where the achievement of programmed milestone dates, cost control/reporting, safety and quality targets are not met. Bespoke clauses have been developed to identify the various cost bases and the addition and deductions to cover performance and payment.

By separate agreement: none.

Operation of the scheme – administration resource needed: a disproportionate amount of time is spent in managing these contracts. This is not a reflection of the contractors' responses but the manner in which the audit of costs requires constant attention to detail. Both the project managers and the contractors consider there has to be a better way to procure such services.

Evidence: views expressed in interviews/documents

Client/consultant: the client believes this type of contract is not achieving the anticipated success. The terms and conditions were developed in-house by the client, since the industry version was not readily acceptable to the client. The strain on working relationships is unsatisfactory and provokes endless argument in on the number of failures and delays and subsequent payments. Where maintenance contracts are separately tendered in contrast to design, build and maintenance contracts then covering the potential defect liability from the original manufacturer is not any easy risk to assess.

Key performance indicators provide the payment release mechanisms based on availability, service hours over a 20 hour period. In general a 98 per cent availability is required during service hours. Negative incentives are applied for a stopped machine (LADs). Positive incentives are applied where 99 per cent availability is achieved.

Contractor: the client's terms and conditions do not reflect the industry's current practice. The contractor wishes to totally manage the assets and be paid on service availability. The type and design of these escalators are unique and do not lend themselves to normal maintenance cycles. Where instant responses are required to attend failures (as defined in the Hong Kong Underground Railway contract) then a different type of maintenance arrangement will apply. This will have a knock on effect to the price. The parameters for success are not embodied in this contract and it frustrates the contractor in trying to provide a reasonable service.

How was the incentive scheme developed and handled: historically, this type of asset has always caused management difficulties in trying to maintain aged machinery. The early target cost contracts eventually became unworkable and the revised performance based specification was developed to improve the situation.

The adversarial nature still persists and causes an inordinate amount of time to manage. Where availability targets can be provided on a realistic basis, then the service maintenance can be incentivised to improve availability wherever technically and financially possible.

Management of the incentive scheme: on-site systems have been developed over the period of the contract to monitor availability in a more sophisticated manner. Historically the client has tried to engineer success within a reducing maintenance budget. Based on this monitoring system, the release of payments can be structured more easily. However it is not promoting a joint partnering arrangement with common objectives.

What improvements could be made on future projects of a similar nature: the development of a centralised procurement strategy is currently under development to enable a wider view to be taken of the difficulties in maintaining a large number of old machines. This approach is actively being encouraged by line managers. The difficulty with very old assets and the renewal of maintenance contracts on a 5–7 year basis is the difficulty in pricing for the risk of failure due to their age and their current capability to perform at the desired level. Whilst the incentivisation scheme does operate in a practical manner, the risks associated with such aged assets mitigates any benefit likely to be derived from the incentivisation of performance. Longer-term design, build and maintenance contracts are likely to replace the current contracts. This rationale will place the design and maintenance risks with the contractor. Performance can then be incentivised.

A significant issue remains however in determining a competitive market with so few contractors manufacturing the size and type of escalators needed to sustain the client's system.

Any additional information for future reference: this area is the subject of PFI review and the Industry would welcome the opportunity to invest in multi-stage escalators more in line with conventional machines in the retail sector. This would require substantial investment in the current infrastructure to accommodate such a change in escalator sizes.

Research contractor's comments
- The setting up of an incentivised contract to improve on performance. Ie improved programmed return dates has been brought about by necessity due to reduced capital renewal programmes. This appears to have had very limited success.

- There are significant unknown risks inherent with the age of some machines. It is not practicable to ascertain the precise extent of the maintenance works until the machine is stripped down for repair. Only then can the full cost be determined. The calculations to reimburse the contractor are workable but cumbersome. The incentivisation scheme is clear but it is not always possible to achieve the time and cost targets. Whilst both parties were focussed on performance, it appears that any improvement on the programmed completion date was the client's true objective but the incentivisation scheme is heavily biased towards costs.

The lessons to be learnt from this type of project
1. Where significant emphasis is placed on incentivising the avoidance of cost and time overruns then any contractor faced with delays due to non-availability of special parts and materials will be under great pressure to seek alternative solutions. The re-use of cannibalised materials from other sites is one way to resolve the problem. Such constant pressure to react to difficult situations with little or no margin for error does not contribute to the alignment of client and contractor objectives.

2. The restricted competition in the marketplace provides little opportunity for any real alternative to provide value-for-money services. The potential for the market to dictate the purchase price is ever present and an alternative procurement strategy will be needed if the spiral of decay is to be halted.

3. The move towards an (DBFO) arrangement would be welcomed by the industry but will require strong negotiating skills by the client to establish tough targets on service availability at an affordable cost.

CASE STUDY 4

Sector: utilities	**Project type:** water treatment plant
Project description: a design and build project to construct a new water treatment plant at a remote location.	

Objective of the incentive scheme: to optimise the performance of the works and obtain the new facility on time and within cost.

Background information: for this particular client it was their first attempt in using an incentivised approach to the construction of new works. They nevertheless wished to retain a tight hold of the programme and costs as for similar traditional contracts. From this experience, they would assess the widespread introduction of incentivised contracts to other projects. Whilst the client's objectives on cost were not precisely met, the target price concept did force all parties to concentrate on performance which for this type of project is the key deliverable as to the water supply which needs to meet the EEC Water Directives.

Incentivisation category: pre-planned	**Overall project value:** £6 million
Overall outcome (success/partial success/failure) Partial success.	**Measurable benefits (time/cost/quality)** **Expected** – two year contract. Within budget. Meet operational parameters. **Achieved** – works completed on time. On target price for mechanical and electrical (M&E) works. Exceeded target price for Civils Works. Operational parameters met.
Form of contract: IChemE "*Green book*".	

Description of scheme documentation: the standard form was heavily modified to reflect client's desire to closely monitor costs throughout project. Following agreement of the target price, the contract was monitored on an open book basis throughout the duration of the contract. The client adopted a very cautious approach, whereby all actual costs were paid with a profit level agreed prior to the commence-ment of the contract. (The net profit was fixed at 3.3 per cent). The contractor's overheads were also kept under review continuously during the construction and commissioning period using a team of auditors. Within the standard form of contract, the clauses were amended to reflect the joint objectives of the client and contractor.

By separate agreement: in addition to the contract agreement, a separate agreement was used to describe how the joint objectives were to be achieved and what needed to be put in place to reflect the partnering arrangements.

Operation of the scheme – administrative resource needed: the client's lack of experience using the incentivised approach caused them to assign more resource than is normal to audit the contractor's costs. This reflected the early stages of developing a mutual trust between the client and contractor, whereby any potential for cost savings within the operational plant area should be declared and shared. The contractor was keen to develop his long-term business relationship with this client and therefore was prepared to go along with their wishes on this first contract.

Evidence: views expressed in interviews/documents

Client/consultant: this design and build contract was assembled on a non-conventional basis with twin targets of M&E works and civil works. The remote location prompted the need to create a close working relationship between the parties if targets were to be achieved. Whilst this was the first project to use an incentivised approach, there were constraints on how far the new approach could be pushed due to resource and cultural changes.

Contractor: a consortium agreement formed the basis of the contract. The M&E works judged a success in terms of time and outturn cost, whilst the civils works were on time but exceeded estimated target cost. The client's objectives were not precisely met, although this was in part due to the broad descriptions used to define their list of requirements. The target price concept was good which encouraged a partnering approach.

How was the incentive scheme developed and handled: to develop the spirit of co-operation, the client considered that a negotiated tendering procedure would produce a meaningful target cost, based on detailed risk assessments discussed between the client and the contractor. This open book stance would be audited throughout the contract so that agreed costs were paid with profit levels also agreed at the beginning of the contract and maintained throughout the payment cycle.

A net profit of 3.3 per cent of the contract sum was agreed by the parties and whilst low in commercial terms, did mean that a profit was made to encourage continuous improvement.

Management of the incentive scheme: with the open book approach adopted throughout the contract period, it focussed more attention on performance rather than variations. The joint management approach relied on the individual personalities and on this contract the personnel from the client and the contractor had worked well on previous contracts. Both parties signed up to the partnering approach as the means of incentivisation through co-operation.

The schedules contained within the IChemE *"Green book"* document included for prescribed tests and performance indicators. This was a key element in the partnering arrangement for progressing the works. On previous contracts, there had been excessive paperwork and it was the agreement of all parties that positive action would be taken to reduce such volumes of unnecessary paperwork. Based on the contractor's experience with similar clients they were able to promote the benefits of partnering and the management reporting cycles which met the client's requirements.

The achievement of withdrawing a tranche of staff out of the management process provided a significant financial benefit to the client.

What improvements could be made on future projects of a similar nature: with the partnering approach, the design phase should initially allow for a moveable cost base. Once fixed, then the agreement on profit levels and overheads will establish the target cost. Any further profit sharing is then agreed on the basis of performance.

These types of projects demand optimal performance of plant and equipment to meet specific requirements in terms of both quantity and quality of water treatment and supply. There is therefore a significant inherent risk within the design process to predict this optimum performance. The client is generally wholly reliant on the expertise of the consultant and contractor to demonstrate that their selection of plant and equipment will meet the client's demands. The client considers that more in-depth risk assessment should take place during the design process, prior to agreeing the target cost.

Any additional information for future reference: more and more clients within the water industry are prepared to develop partnering arrangements to incentivise the contracts so that the win/win situation occurs. Predictability of achieving the performance specification to meet EEC Directives is now key to their success and this will encourage a more flexible style of contract. Further standardisation of the most appropriate contract form will assist clients to incentivise performance.

Research contractor's comments

- The use of an incentivised contract to improve on performance as opposed to using the traditional contract forms MF1 and/or ICE6th was seen as an experiment to determine the benefits for the client. In this case study with the M&E contractor as the main contractor and the civils contractor as the sub-contractor, there clearly was little opportunity to recover the exceeded target cost on the civils work.

- There are significant design risks in water treatment projects and this issue is generally agreed throughout the industry. Therefore to properly quantify these risks during the design phase and then agree a target cost would appear to be a better basis upon which to focus attention on achieving optimum performance. The contractor would then tackle the problems knowing that any increased costs would not all be to his account. The calculations to reimburse the contractor are workable but cumbersome.

- The incentivisation scheme is clear but it is not always possible to achieve the time and cost targets. Both parties were focussed on meeting the plant performance targets during the commissioning stage, but the incentivisation scheme is heavily biased towards meeting the target costs.

The lessons to be learnt from this type of project

1. Where significant emphasis is placed on incentivising the avoidance of increased cost due to design risk then any contractor faced with delays due to non-achievement of plant and equipment performance during commissioning will be under great pressure to seek alternative solutions. Such pressure to react to difficult situations with little or no margin for error does not contribute to the alignment of client and contractor objectives.

2. The level of knowledge on new plant and equipment within the marketplace is held by a relatively small number of suppliers. Until such new plant is physically working in the field, then the client has little alternative but to accept bench test results. The potential for the market to dictate the purchase price is ever present and an alternative procurement strategy will be needed if the target cost approach is to remain meaningful and a practicable way to manage such projects.

CASE STUDY 5

Sector: utilities	**Project type:** sludge drier
Project description: this was a new facility built onto an existing sewage treatment works. This type of facility is relatively new technology within UK and there was little or no experience within the client's organisation to manage the commissioning of such technology. The engagement of a consultant to project-manage the construction and commissioning was therefore seen as a prudent use of resources to minimise the risk of introducing new technology plant.	

Objective of the incentive scheme: to meet the programmed operational date and to contain costs within the client's budget.

Incentivisation category: pre-planned	**Overall project value:** £1.1 million
Overall outcome (success/partial success/failure) Failure. In terms of time and cost this project has failed on both counts. As to quality, the current commissioning trials are expected to achieve the desired performance levels. The client considers the inherent risks in introducing the new technology plant needed more assessment. In hindsight the target price appears to have been too optimistic.	**Measurable benefits (time/cost/quality)** **Expected** – complete by July 1997. Within budget. **Achieved** – well over time. Commissioned in February 1998 Cost including variations has escalated to £1.6 million
Form of contract: IChemE *"Green book"*.	

Description of scheme documentation: the standard form was heavily modified to reflect the client's desire to closely monitor costs throughout the project. Following receipt of tenders, negotiations were held to determine the target cost. As this form of contract requires, the clauses were amended to reflect the project objectives in terms of time and cost. Liquidated and ascertained damages were also agreed for time and operational performance overruns. The target cost was made up of the negotiated elements of engineering management, materials and design, with the overheads and profit as fixed percentages of the negotiated elements. There was a shared risk on target prices, but the non-target prices were at the contractor's risk.

The incentivisation of the contract was aimed towards improved performance, covered the opportunity to make design changes with cost savings which formed part of the overall shared savings scheme.

Following agreement of the target price, the contract was monitored on an open book basis throughout the duration of the contract. The client adopted a very cautious approach and the contractor's costs were kept under continuous review during the construction and commissioning period.

By separate agreement: none.

Operation of the scheme – administrative resource needed: the client recognised their lack of experience in the use of the incentivised contract approach and in the introduction of new technology plant. They considered that to compensate for this shortfall more management resource should apply than as in the case of the traditional approach using a MF1 and ICE 6th forms. The reliance on others external to the client to manage the project and the use of the internal Engineering Services Team acting as the client's representative, caused a heavy administrative overhead to be applied to the target cost.

Evidence: views expressed in interviews/documents

Client/consultant: since other water companies were utilising the incentivised approach with some success, the client considered that it should be trialled on this project to ascertain its potential use on future projects.

To properly focus the introduction of the new technology, the contract was tendered on the basis of using the M&E contractor as the lead contractor with several subcontractors for civil and other works. This was set up on a non-integrated Joint Venture basis. The client stated that they may have benefited from more hands-on involvement rather than leave matters to be resolved by a third party. This created a "blame culture" and although all the contract arrangements were not wholly tested, there was a lack of clarity in implementing a partnering approach.

The introduction of new technology plant required a better understanding of the hand-over process during commissioning. This would have helped considerably in reducing the tensions between the parties.

Contractor: although the client had not used this form of contract before and the introduction of a target cost contract with incentives was under trial, the main contractor was familiar with its workings and had operated such an approach with other clients throughout the UK and was aware of the issues that could arise. They felt that this experience would be a positive contribution to the trial project. With two nominated sub-contractors working with a civils contractor, the main contractor considered that the management of the works required greater intervention by the consultant/QS than anticipated. The main contractor considered that the potential for repeat business provided an added incentive to ensure that this contract was completed within the agreed targets.

How was the incentive scheme developed and handled: whilst the client considered this approach was an act of faith, the open book accounting process gave them appropriate mechanisms in which to monitor and control the contract. The client considered that a negotiated tendering procedure would produce a meaningful target cost, based on detailed risk assessments discussed between the client and the contractor. This open book stance allowed the contract to be audited throughout the contract period so that agreed costs were paid with profit levels also agreed at the beginning of the contract and maintained throughout the payment cycle. A net profit of 3.3 per cent of the contract sum, whilst low in commercial terms did mean that a profit was made to encourage continuous improvement.

Management of the incentive scheme: with the open book approach adopted throughout the contract period, it focussed more attention on performance rather than variations. The joint management approach relied on the individual personalities and on this contract the parties from the client and the contractor had worked well on previous contracts. Both parties signed up to the partnering approach as the means of incentivisation through co-operation. The schedules contained within the *"Green book"* document included for prescribed tests and performance indicators. This was a key element in the partnering arrangement for progressing the works. On previous contracts, there had been excessive paperwork and it was the agreed intent from this experience not to perpetrate such volumes of unnecessary paperwork. Based on the contractor's experience with similar clients, they were able to promote the benefits of partnering and the management reporting cycles which met the client's requirements.

What improvements could be made on future projects of a similar nature: with the partnering approach, the design phase should allow for a moveable cost base. Once fixed, the agreement on profit levels and overheads will establish the target cost. Any further profit sharing is agreed as an incentive to the partnering arrangement.

Any additional information for future reference: more and more clients within the water industry are prepared to develop partnering arrangements to incentivise the contracts so that the win/win situation occurs. Predictability is now key to their success and this will encourage a more flexible style of contract. The development of model incentivisation schemes within this sector would be welcomed.

Research contractor's comments
- The use of an incentivised contract to improve on performance as opposed to using the traditional contract forms MF1 and/or ICE 6th was seen as an experiment to determine the benefits for the client. In this case study with the M&E contractor as the main contractor and the civils contractor as a sub-contractor, there clearly was a greater need to manage the interfaces between contractors as part of the incentivisation scheme.

- There are significant commissioning risks in introducing new technology. Thus to properly quantify these risks during the design phase and then agree a target cost would appear to be a more informed approach to managing performance. Where it is not practicable to ascertain the precise extent of the risk, then some sharing mechanism should be applied. The contractor would then tackle the problems knowing that not all the increased costs would be to his account. The calculations to reimburse the contractor are workable but cumbersome. The incentivisation scheme was clear but this does not appear to have assisted in the resolution of problems. Both parties were focussed on meeting the plant performance targets during the commissioning stage, but there seems to have been insufficient emphasis placed on the wider issues surrounding the incentivisation scheme.

The lessons to be learnt from this type of project
1. Where significant emphasis is placed on incentivising the avoidance of increased cost due to design risk then any contractor faced with delays due to non-achievement of performance tests during commissioning will be under great pressure to seek alternative solutions. Such pressure to react to difficult situations with little or no margin for error does not contribute to the alignment of client and contractor objectives.

2. The level of knowledge on new plant and equipment within the marketplace is held by a relatively small number of suppliers. Until such new plant is physically working in the field, then the client has little alternative but to accept bench test results. The potential for the market to dictate the purchase price is ever present and an alternative procurement strategy will be needed if the target cost approach is to remain meaningful and a practicable way to manage such projects.

CASE STUDY 6

Sector: utilities	**Project type:** water treatment works extension
Project description: the construction of an extension to the existing works at a remote site.	

Objective of the incentive scheme: to ensure successful completion of the project within clearly defined cost and programme parameters.

Background information: based on the experiences of other utility companies, this project was promoted on the extended arm contracting principles. It was considered a pathfinder project for this client who wished to develop the opportunities of the incentivised approach in meeting capital spending goals.

Incentivisation category: pre-planned	**Overall project value:** £1.9 million
Overall outcome (success/partial success/failure) Success.	**Measurable benefits (time/cost/quality)** **Expected** – 24 months including the design phase and within budget. **Achieved** – built to time and to budget including variations.
Form of contract: IChemE *"Green book"*.	

Description of scheme documentation: heavily modified to reflect an extended arm contracting arrangement. A series of sub-contracts let by client under the *"Green book"* form. No contractual arrangement between sub-contractors but a loose partnership arrangement agreeing to work jointly together with the client to achieve common goals. Based on a client brief, the brief gave an individual contract statement of need. The brief also gave an indicative target cost, an anticipated timescale within which to complete the scope of works and identified the client's management team. The client employed the contractor with the design team, to manage the works for a fee and employed discrete trades contractors to carry out the works.

By separate agreement: none.

Operation of the scheme – administrative resource needed: as a pathfinder for future use this project could have been operated more smoothly. With new concepts of managing and controlling time and cost, there was the inevitable learning curve to climb. However the client's belief in the new approach for the contractor to manage more effectively the downstream problems, within the team, led to more resource than normal being applied. With a remote location, staffing was a constant issue which had to be managed.

Evidence: views expressed in interviews/documents

Client/consultant: a form of management contracting, the client believed in the opportunities afforded by the *"Green book"* conditions rather than the use of ICE 6th edition. The concept was good and offered a sound way forward on a remote site.

The overall net benefit had to be viewed as the difference between the pressure to minimise outturn costs and the learning curve of the staff operating the contract.

The role of the commissioning engineer proved a sound investment in signing off milestone achievements.

Contractor: this contract was similar in nature to previous contracts with another client. The reasons for the success on this particular contract relates more to the effort put in to managing a contract at the opposite end of the country and resolving the difficulties of a remote site.

Lack of knowledge of the local sub-contractors led to misunderstandings with these sub-contractors as to the operation of the extended arm approach.

The success of this type of loose partnership arrangement relies heavily on the individuals involved.

How was the incentive scheme developed and handled: the incentivisation was handled by negotiations in a friendly manner and openness at Board Level. There were differences of opinion as to how the partnering approach should be adopted. Proposals submitted by the sub-contractors were developed to a final schedule of rates with a design fee. Agreement to a target cost with an equitable gainshare/painshare arrangement provided the basis for minimisation of outturn cost.

The setting up was relatively straightforward but the implementation proved difficult due to remoteness of the site.

Management of the incentive scheme: the extended arm approach placed the team building emphasis equally amongst the parties. The management of downstream problem solving protected the client's interests. It focussed more attention on performance rather than seeking variations.

The joint management approach relied on the individual personalities and on this contract all parties worked well together. The client considered that the sub-contractors were on their side.

The management style and approach was carefully orchestrated and the project was completed on time and to cost including all variations. The plant is operating satisfactorily.

Based on similar experience with other clients, the sub-contractors were able to promote the benefits of partnering and the management reporting cycles which met the client's requirements.

In such circumstances the role reversal of "poacher turned gamekeeper" helped to promote understanding on both sides of the partnering arrangement.

What improvements could be made on future projects of a similar nature: whilst resources were adequate on the subcontracting side, more experience of this type of arrangement would have helped. The remote location of the site caused staffing difficulties. This is significant when developing the total team concept.

The client team needs to have a clear coordination and facilitation brief to fully develop this type of incentivisation through partnering. Having a vertical management structure with non-matrix project managers can lead to weaknesses in applying this approach.

Any additional information for future reference: more and more clients within the water industry are prepared to develop partnering arrangements to incentivise the contracts so that the win/win situation occurs. Predictability is now key to their success and this will encourage a more flexible style of contract.

The key to successful delivery of such schemes is the people involved. The staff with the right attitude and approach will ensure the win/win situation is achieved. Those staff who set out to be over-bearing and demanding will cause the team approach to unravel and put pressure on the project's outcome.

Research contractor's comments: the client had recognised that a strong project team of known personalities would promote the right message to the sub-contractors. The client's learning curve appeared to be a sound investment in time to introduce a new way of operating to improve performance. With the operating environment well thought out, it required all project team members to work together to achieve common goals.

The concept offered the client a good way forward. The lessons learnt from the implementation will be reinvested into future projects, where the learning curve will be minimal and the value-for-money will be greater on projects of a similar nature.

CASE STUDY 7

Sector: utilities	Project type: water ring main improvements
Project description: the construction of a major pipelines project designed to improve the security and flexibility of supply to customers.	

Objective of the incentive scheme: to achieve a considerably faster construction time than a conventional distribution mains installation and for at least part of the works to be finished "as soon as possible" to deliver the benefits of the scheme before the summer demand.

Background information: following on from a " partnering arrangement" between the client and the contractor, the team included a designer and a further specialist contractor to provide additional construction resource. The client subsequently engaged a construction consultant to provide a project management and CDM support service to the project team including development of a contract strategy.

Incentivisation category: pre-planned	Overall project value: £7.5 million
Overall outcome (success/partial success/failure) Success.	**Measurable benefits (time/cost/quality)** **Expected:** to meet the 31 March and 30 May 1997 dates within budgeted costs. **Achieved:** both dates achieved within the target costs.
Form of contract: IChemE *"Green book"*.	

Description of scheme documentation: heavily modified to reflect an extended arm contracting arrangement. A series of sub-contracts let by the client under the *"green book"* form. In line with the partnering principle and especially in view of the mutual interdependence of each member of the project team, it is apparently equitable for each contractor to be reimbursed according to their input. This form of contract is already known to the parties. As a result, extensive data is available for comparative purposes to enable prediction of outturn costs. The agreement to operate a target cost for each contract prior to works commencing on site was based on an agreement of the predicted final outturn costs. Once the target cost was agreed, individual contracts and hence the project were managed within this target. It maximised the incentive for all members of the project team to reduce actual construction costs in the field and was a powerful lever towards problem resolution.

By separate agreement: no contractual arrangement between sub-contractors but an agreed mission statement signed by all parties to work jointly together with the client to achieve common goals.

Operation of the scheme – administrative resource needed: for the avoidance of doubt it is noted that the term "partnering" in the context of this project does not imply any intent whatsoever by the parties to create a legally binding business partnership or joint venture. Conventional arguments against "cost plus" include the complexity (and cost) of the payment application process. By developing and sharing existing systems already in use by the main contractor, this feature was minimised and indeed that assistance to the establishment of the overall cost system was self-financing.

Evidence: views expressed in interviews/documents

Client/consultant: a form of management contracting without a "main contractor" so as not to upset the "partnering concept". The client considered that the adoption of the target cost approach arising from the IChemE *"Green book"* form gave ample incentives to all parties. There were considerable time pressures on availability of the new mains to meet planned closures of other facilities. Also the vulnerability of the network to sustain a supply without the new works commissioned. Need for project to proceed considerably faster than a conventional installation. Time was of the essence.

No formal contracts exist between the contractors. The partnering arrangements comprised of distinct contracts between the client and each party involved.

Partnering is intended to describe the management approach or philosophy adopted to achieve the parties' specific business goals by establishing mutual objectives, with measures in place for non-adversarial problem resolution and continuous improvement. A mission statement was signed by all parties. This stated the key objectives and the manner in which this was to be achieved. A responsibility matrix clearly identified the contributions from each party to the key tasks.

Contractor: all sub-contractors experienced in this type of arrangement. The arrangements provided for an all-embracing service of management, construction and design skills, to meet the diverse and expedient needs of the client. This project was similar in nature to previous projects with the client. The reasons for the success relates more to the desire to continue to carry out further projects which arise.

Key staff forged exceptional working relationships with all parties and performance is the focus of attention. The success of this type of loose partnership arrangement relies heavily on the individuals involved. For the avoidance of doubt, the term "partnering" in the context of this project does not imply any intent whatsoever by the parties to create a legally binding business partnership or Joint Venture. The true incentive is to continue to meet the client's individual project objectives and seek the opportunity to carry out repeat business.

How was the incentive scheme developed and handled: the incentivisation was handled by negotiations in a friendly manner and openness at Board Level. There was a clear understanding of the roles and responsibilities and the mutual interdependence of each member of the Project team, it was equitable for each contractor to be reimbursed according to their input.

Extensive data was available for comparative purposes to enable the prediction of outturn costs. The agreement to operate on a cost reimbursable arrangement enabled the earliest possible formalisation of works contracts. It maximised the incentive for all members of the project team, including the client, to reduce actual construction costs in the field and provided a powerful lever towards problem resolution. It also minimised the potential for dispute arising from weather, unforseen conditions and enabled the team to concentrate on getting the job done.

Conventional arguments against "cost plus" include the complexity (and cost) of the payment application process. By developing and sharing existing systems in use, this feature was minimised and indeed that assistance to the establishment of the overall cost system was self financing.

Management of the incentive scheme: the extended arm approach placed the team building emphasis equally amongst the parties. The management of downstream problem solving protected the client's interests. It focussed more attention on performance rather than seeking variations. The joint management approach relied on the individual personalities and on this contract all parties worked well together. The client considered that the sub-contractors were on their side.

Based on previous experience with the client, the parties were able to promote the benefits of partnering and the management reporting cycles to meet the client's requirements. The commitment to the project objectives and an equitable share in the gain or loss arising from its outcome was the essence of the partnering approach. Time related bonuses were attached to target dates payable in full if they were achieved and not at all if dates were missed, whatsoever the reasons.

The cost target was calculated on the basis of agreed resource predictions, verified by reference to historical client data and other references. At the end of the project, the total variance from each of the two cost target elements was calculated. 50 per cent of this variance was assigned to each of the parties subject to a maximum limit of 25 per cent of the respective cost target element being the maximum total credit.

What improvements could be made on future projects of a similar nature: this project is clearly a benchmark for incentivisation within a partnering environment. Its applicability within the utilities sector is generally accepted. Refinement for other sectors may be required.

Any additional information for future reference: the key to successful delivery of such schemes is the people involved. The staff with the right attitude and approach will ensure the win/win situation is achieved. Those staff who set out to be over-bearing and demanding will cause the team approach to unravel and put pressure on the project's outcome. Board level commitment is clearly a critical success factor when using an incentivisation approach.

Research contractor's comments: ideally, to optimise progress and co-operation, the project team should operate from one location. On this project, with design work well advanced, the disruption to the process by relocation to the construction site was not warranted. Instead, the design element of the team remained at their offices with frequent daily contact with the construction/contractors part of the team during the pre-construction period. The minimisation of outturn cost as a project objective was supported by an equitable "gainshare/painshare" arrangement based upon a target cost provided for within the contracts.

"Partnering" was intended to describe a management approach or philosophy which was adopted to achieve the parties specific business goals by establishing mutual objectives, with measures in place for non-adversarial problem resolution and continuous improvement. This approach was deemed successful as all parties received a benefit from it through increased co-operation, lack of confrontation and elimination of litigation enabling maximum focus on the project objective.

CASE STUDY 8

Sector: transportation – public sector	**Project type:** renewal and refurbishment of buildings and structures associated with a passenger railway
Project description: to carry out minor and major works to buildings using an extended arm contract approach which affords early mobilisation of the contractors over conventional tendering procedures.	

Objective of the incentive scheme: to encourage a value for money approach and meet exacting annual capital programme targets.

Background information: using a conventional tendering approach, the timescales to mobilise the works would not allow sufficient works to be completed within the annual spend profiles. By selecting three contractors using a detailed pre-qualification methodology, the opportunity exists to fast track the volume of works.

Incentivisation category: pre-planned	**Overall project value:** varies annually between £20–30 million
Overall outcome (success/partial success/failure) Success.	**Measurable benefits (time/cost/quality)** **Expected** – for each project to be completed to time and within the target cost. In addition special emphasis is placed on maintaining the safety of the railway. **Achieved** – Whilst some individual projects had time overruns, this has not caused the overall programme targets to be missed. Generally, the good performance of the contractors has led to a small under-expenditure on the programme. Quality and safety targets were achieved.
Form of contract: hybrid JCT 80 into a Framework Agreement.	

Description of scheme documentation: the framework agreement sets out the prime cost and target cost relationships, their definition and implementation. It describes a two part contract, where the first part is the reimbursable contractor's costs and the second part relates to tendered sub-contracts for each project released under the agreement. Specific clauses have been included to highlight the methodology and application.

By separate agreement: none.

Operation of the scheme – administrative resource needed: there are a significant number of projects which have to be managed. The client's team and the contractors project team all work in the same office and this has allowed more dialogue than is the case on a conventional contract. Direct action has lead to streamlined procedures whereby the client is able to keep to a minimum their contract monitoring staff.

Evidence: views expressed in interviews/documents

Client/consultant: extended Arm Contracts (EAC) have the overall aim of the ability to mobilise projects as quickly as possible. Each project is completely tendered with an A4 sheet describing indicative rates, including overheads and profit. There are three contractors on the framework list, each with a design and build capability. Each project released, goes through a feasibility stage, then onto a design stage with cost estimate. This covers the contractor's costs, prime costs and the client's costs to establish the Project Expenditure Summary.

This type of agreement enables a large number of lower value projects to be expedited. The current programmes are on target to achieve the level of spend.

Contractor: the framework agreement provides a continuous source of work. To be placed on this agreement has meant going through a tendering exercise to agree rates for application to management tasks.

This form of agreement extends the railway knowledge to project and works level, where the partnering roles are formed to achieve common goals. There is one project office where the team (the client and contractors) work alongside one another. Safety and quality are paramount. Each framework contractor manages the sub-contractors on each project. There are financial parameters set to invoke competitive tendering procedures for the subcontractor's prices.

How was the incentive scheme developed and handled: where projects are brought in under budget then an incentive payment is calculated in accordance with the Framework Agreement and paid to the contractor.

Performance monitoring is the key to successful implementation, with cost control/reporting; safety management; quality management and achievement of milestone dates as key indicators.

Audit checks are carried out to review contractor costs on an open book basis. In general, time overruns are not penalised but a keen check is maintained to monitor expenditure. Variations and change in scope have to be authorised prior to commencement, otherwise no payment will be made.

Management of the incentive scheme: the target cost is the projected cost of the works as agreed between the parties. In the event of the following circumstances a bonus will become payable to the contractor. Where the final account of the works is less than the finally adjusted target cost by more than 5 per cent of the finally adjusted target cost then a payment will become payable to the contractor. The amount of any payment due for the target cost adjustment is calculated using an agreed formula subject to an upper limit of £150 000. Reaching this threshold will be subject to audit prior to payment.

Abatement provisions apply where the contractor fails to achieve satisfactory levels of performances described above. This includes data input to the client's project management system by due dates. In these circumstances, deductions are made from the contractor's invoice in accordance with an agreed scale. Using the overall incentive approach, the opportunity exists to price and commence new projects easily.

What improvements could be made on future projects of a similar nature: to make better use of this approach requires a more flexible budgeting approach in the public sector. Approaching year-end, if the contractor has been successful in reducing the overall expenditure for the total amount of works then this puts pressure on the client to explain the under-expenditure. Thus the benefits of reduced costs are not evenly distributed. Within the Governments annual accounting arrangements there is generally a lack of understanding of EAC. The controls needed to enforce the management regime are cumbersome.

Any additional information for future reference: where annualised programmes exist within the public sector, there are sometimes great difficulties in spending the budgeted shortfall before the end of the financial year end. There is in effect a positive disincentive for the client not to balance the books since it is a lost opportunity to spend the money. There is generally no carry over from one year to the next. Freedom to operate within a more flexible funding arrangement would enhance the effectiveness and efficiency of EAC.

Research contractor's comments: the purpose of EAC is to mobilise works on a fast track basis, within a management framework to meet the targets of time, cost, quality and safety. The time and cost targets do not always sit comfortably together. The relaxation of annual budgets to three-year spending targets will allow a more sensible carry-over of projects from one year to the next. However at the end of each three-year programme, there will still be a race to finish projects on programme to spend the financial allocations. This effect will not be as dramatic as before. Provided the contractors meet their targets for each project then any shortfall in the programme will be a lost opportunity to both parties. This is a shared risk. These types of contract carry high management overheads and it is imperative to both parties to maintain a workload to justify this cost. The size of the annual workload is therefore critical so that the contractors can recover their costs. The contractors are actively encouraged to seek out improved operating methods and techniques. There is a direct correlation between cost savings and bonus payments, provided quality and safety criteria have been met.

CASE STUDY 9

Sector: transportation – public sector	Project type: roads – motorway-widening schemes
Project description: to increase the number of lanes on both carriageways to meet the increasing demands whilst maintaining safe traffic flows at all times.	

Background information: The motorway improvement schemes are symptomatic of the ever-increasing car and HGV traffic demands. To keep congestion to a minimum, the construction programme must encourage the contractor to minimise delay by using innovative solutions to control traffic movements and safeguard the workforce. The use of pre-cast concrete barriers as a protection method allows greater vehicle passing speeds than normal. The prime objective is to complete the works as quickly as possible and free up the lane to ease traffic flows as soon as possible. Any potential cost saving will be recognised by the client when awarding future contracts.

Incentivisation category: pre-planned	Overall project value: varies but generally £50–200 million
Overall outcome (success/partial success/failure Success.	Measurable benefits (time/cost/quality) Expected – to significantly improve on the time and cost overruns on current projects, typically up to 38 per cent cost overrun. Achieved – contained within a band of 5–10 per cent variation over the life of the contract. The majority of contracts are on time with some early finishes.
Form of contract: ICE 5th Edition.	

Description of scheme documentation: Design and Build Contract with Clause 12 (unforeseen circumstances) left in. The schedule of milestone payments is contained within a minimum and maximum contract period. In addition there are modified clauses to control payment and performance.

By separate agreement: none.

Operation of the scheme – administrative resource needed: with the incentive to manage the work on time, the contractor is actively encouraged to invest in sufficient resource to maintain the target dates. The build up of trust with the client was seen as paramount when problems arose and speedy resolution were required. Monitoring quality more closely means a better job with the same time and cost parameters.

Evidence: views expressed in interviews/documents

Client/consultant: this project was completed on time with claims contained within a 5–10 per cent variation of the tender price. Client changes were made but these were kept to a minimum following experience on other contracts which gave rise to significant cost overruns.

Where the contractor suggests changes which do not reduce the quality of the product, then all savings will go to the contractor. Where changes are suggested which reduces the quality, there will be a compensating reduction in the tender sum. There is the incentive for the contractor to programme work to beat the target time. Design risk is also passed to the contractor.

Contractor: with the down turn in other construction sectors, the pricing for motorway work became keener. This required the contractor to look at working methods again and to promote more radical technical solutions. Wholesale closure of lanes for extended periods was not acceptable and any piecemeal refurbishment programme meant high cost and low returns. The transfer of certain risks to the contractor is based on an agreement with the client that those risks are best managed by the contractor.

How was the incentive scheme developed and handled: based on past experience with similar schemes, it was paramount to incentivise the contract to encourage an early completion if possible. By packaging the design and construction risks together, the opportunity was taken to minimise the level of claims. This was linked to lane rental agreement which allowed occupation on an agreed basis.

Management of the incentive scheme: traditional milestone payments with the performance monitored as to quality and quantity. Lane rental overruns would be costed and charged to the contractor.

What improvements could be made on future projects of a similar nature: none. The new approach is working well. Likely to become the benchmark, although the new ECC may replace the use of ICE 5th.

The development of partnering will introduce arrangements to identify problems in advance. Use of value engineering techniques will be included within the formal agreement. Any savings will be shared.

Any additional information for future reference: greater trust to be developed within the industry to reduce the adversarial content. In the long run, clients may pay more for the jobs done but with less claims or disputes and consequently overall there is likely to be a reduction in costs.

Research contractor's comments: with increased traffic demands, the urgency is on the client to develop more sophisticated contract arrangements to meet the new time scales. In addition the use of the ECC will foster the partnering style. We understand that reasonable progress is being made on this front. In respect of introducing new management techniques, the use of value engineering as part of the formal agreement has resulted in better schemes and cost savings. Where the contractor proposes real savings without loss of quality, they directly benefit. Alignment of the strategic objectives has brought significant financial rewards to both parties.

CASE STUDY 10

Sector: transportation – public sector	Project type: roads – motorway maintenance schemes
Project description: to repair the structural and carriageway faults on various motorways using a lane rental arrangement.	

Objective of the incentive scheme: to reward innovation in design of modifications and repair to structures and encourage the use of lane rentals for carriageway repairs.

Background information: to minimise the disruption to traffic it is paramount to keep lane restrictions to a minimum. This can be brought about by the use of lane rentals, whereby ownership of the lane(s), effectively transfers to the contractor for the rental period. In this way the contractor is encouraged to keep control of the use of the lanes and to maximise his working time within the rental period. In respect of structural repairs, there is generally no time restriction since in general the works do not interfere with traffic flows It is more difficult to obtain competitive prices for these works and therefore the client includes the works within the Lane Rental schemes wherever possible. However, the client does encourage the use of innovative techniques to repair the structure. This will impact on overall costs and bring added benefit to the client.

Incentivisation category: pre-planned	Overall project value: various, generally about £2 million per scheme
Overall outcome (success/partial success/failure) Partial success.	Measurable benefits (time/cost/quality) Expected: each series of repairs has an agreed target cost with a bonus for early completion. Achieved: each scheme is technically challenging and they do not all succeed in meeting or beating targets, but there has been an improvement compared to previous contracts.
Form of contract: ICE 5th edition – Re-measure.	

Description of scheme documentation: hybrid document to cover structural repairs and roadworks. Clause 12 (unforeseen circumstances) left in. Heavily modified to reflect differing nature of structural repairs and roadworks.

By separate agreement: none.

Operation of the scheme – administrative resource needed: with the incentive to manage the work on time, the contractor is actively encouraged to invest in sufficient resource to maintain the target dates. The build up of trust with the client was seen as paramount when problems arose and speedy resolution were required. Monitoring quality more closely means a better job with the same time and cost parameters.

Evidence: views expressed in interviews/documents

Client/consultant: to promote value for money, it has been necessary to include the structural repair works as part of the carriageway repairs schemes wherever possible. There are few contractors with the capability to do the structural repair works. Due to the complexity of this work, time targets remain largely inappropriate. They are generally carried out without restrictions of the traffic. There is significant variation in the extent and type of repairs to any structure, which requires the payment for plant and labour, subject to rigorous inspection. Payment is made on the basis of quantities and rates. A major problem is obtaining appropriate documentation to provide defects liability cover greater than two years. There are benefits to both client and contractor to have joint use of plant and labour for the two schemes side by side. Experience has proved that letting structural repair contracts by themselves is difficult. For roadworks contracts there is the real incentive of lane rental costs. Bonuses are pitched to be included within the contract price.

Contractor: these hybrid contracts demand contractors with a broad range of maintenance and repair skills. The key skill is the mobilisation of resources to respond to changing circumstances, particularly ensuring that roadworks are carried out with the minimum of disruption. The risk and reward of this type of contract is heavily dependent upon monitoring progress very closely. The supervisory role and liaison with the client meant that communication of problems and issues was handled quickly to ensure maximum utilisation of the lanes. Wherever possible, innovative methods were deployed to reduce occupation of lanes which correspondingly reduced rental costs. To meet time targets, it is advantageous to the contractor to flood the site with resources.

How was the incentive scheme developed and handled: based on past experience with similar schemes, it was paramount to incentivise the roadworks contract to encourage an early completion if possible. By packaging the design and construction risks together, the opportunity was taken to minimise the level of claims . This was linked to lane rental agreement which allowed occupation on an agreed basis.

Management of the incentive scheme: the contractor was encouraged to innovate throughout the contract term. The traditional milestone payments exist with the performance monitored as to quality and quantity. Lane rental overruns would be costed and charged to the contractor. The mix and match of different types of work was favoured by the client to ensure unattractive work was done within the profitable work.

What improvements could be made on future projects of a similar nature: none. The new approach is working well. Likely to become the benchmark, although the new ECC may replace the use of ICE 5th.

Any additional information for future reference: LADs as negative incentives are best left out of highways contracts.

Research contractor's comments: the expression "there had to be a better way" sums up the success of using incentives to road repairs schemes. The inclusion of structural repairs is a contract strategy due to the difficulties of the marketplace. Where the risks have been well defined and allocated to the appropriate party, then performance can be incentivised. Where jobs take on a greater size, then the opportunity to apply bigger bonuses works well. There appears to be a much better alignment of client and contractor objectives on this new approach.

CASE STUDY 11

Sector: heavy process	Project type: new power station
Project description: the construction of a new power station on a former power station site.	

Objective of the incentive scheme: the introduction of the scheme was an attempt to recover lost time on the heavy civil works by accelerating the works to enable the mechanical and electrical services to keep on target and meet the "on stream" date.

Background information: this contract was a turn-key project which needed to be completed by the due date to supply power to the national grid. The supply contract carried very heavy penalties if the due date was missed which in turn was reflected in the construction and equipping contracts. The new station was constructed on the site of a former power station and preliminary groundworks revealed extensive contamin-ation by asbestos. This should have been removed from the site under the previous demolition contract and was not revealed during routine site investigation. Once discovered, this had to be removed by special licence before the new foundations could be constructed. It was at this point that incentives were considered to try to recover the lost time.

Incentivisation category: reactive	Overall project value: £20 million
Overall outcome (success/partial success/failure) Failure.	**Measurable benefits (time/cost/quality)** **Expected** – originally a 24-month contract, the introduction of £225 000 acceleration payments were set up to encourage the civils contractor to work faster and meet revised deadlines. **Achieved** – despite their best efforts, it was not possible for the civils contractor to fully recover the lost time. Overall the project was two months late on the civils work and five months late on the turbine installation and the outstanding claims are the subject of litigation.
Form of contract: ICE 6th edition, fully designed scheme.	

Description of scheme documentation: none expressed within the original contract other than the significance of the lads to cover time overruns.

By separate agreement: none.

Operation of the scheme – administrative resource needed: with time pressures mounting, the client believed that offering incentives to accelerate the construction programme would resolve the problem. However, the problem was deep seated within the physical elements of the demolished site. Therefore, the scheme was aimed at galvanising the contractor to new levels of resourcefulness which in turn would lead to higher output. This was just not possible within the construction constraints.

The amount of administration required to manage the bonus payments was absorbed within the overall client management cost. Time being of the essence, the lads were significant. Every conceivable opportunity was taken jointly by the parties to accelerate the works.

Evidence: views expressed in interviews/documents

Client/consultant: to take the opportunities which arose out of the deregulation of the electricity supply industry, the client used their extensive knowledge of the generation industry to build new power stations and supply electricity to the national grid. The use of the turnkey approach meant that the client could specify the power demands and then engage a large specialist contractor to oversee the design and construction.

Contractor: with litigation pending, it is not hard to pinpoint the root cause of the time overruns. The expensive removal of the asbestos also increased the contractor's costs significantly. The civils contractor considered that they should have been given the lead contractor role so that they could have led the design team as well as provide specific project management skills. The nomination of the electrical contractor as the lead contractor was seen initially as the most appropriate way to manage the building and equipping programme.

The site had outline planning permission, but needed a detailed consent to comply with planning regulations prior to the commencement of construction.

The overall contract strategy was to complete the structural works, install the large turbines and finally complete the services. In hindsight the risks associated with building a new structure on a previous station site, should have been more carefully evaluated. The amount of site investigation was seriously under-estimated. The majority of the subsequent problems stemmed from this weakness. Without a strong parent company to back the contractor, the project would have failed.

When projects start to go wrong, there is a tendency to throw money at the project to recover the situation. When faced with a series of events which threaten the success of the project, it is encumbent on all parties to try and remedy the situation. For their part the civils contractor instigated interim completion dates to facilitate plant access. This still did not make the desired impact and eventually they were forced into the inevitable time overrun.

People issues are very important when the going gets tough. The working relationship with the client was strained which did not help. In addition design information was arriving late and with the difficulties in communicating with the client, problems which arose were not easily sorted out.

Every resource was focussed to try to achieve the revised dates. Trying to achieve the impossible cannot be incentivised. Targets have to be realistic.

How was the incentive scheme developed and handled: as the name implies, the reactive incentive scheme was set up to offer additional payments if revised targets could be met. Initially the revised targets were formulated by the lead contractor who clearly did not fully comprehend the implications of the extreme construction difficulties being encountered. Once the extent of the delays had been understood then the revised programme dates were issued. The achievement of these dates would attract the bonus payment but the failure to achieve an initial stage did not preclude the achievement of a bonus on a subsequent stage.

Management of the incentive scheme: record keeping was of paramount importance to ensure that precise explanations for delays were recorded. The achievement of some key handover dates meant that some bonus payments could be made.

What improvements could be made on future projects of a similar nature: the absence of a detailed site investigation report meant that the overall target completion date was put at risk. The commercial arrangements could possibly be handled differently in that the programmed handover date should have been linked to a risk sharing arrangement whereby unforeseen risk is shared. Large turnkey contracts need to be managed by the most experienced contractor with sufficient key skills.

Any additional information for future reference: this project reinforces the fact that the costs of an appropriate site investigation are never wasted and are an investment towards minimising unforeseen circumstances. Any project of such a magnitude could not suffer the seriously delayed start and recover the position. The concept of partnering was not considered and maybe such an arrangement may have forced the renegotiation of terms with the national grid.

Research contractor's comments: with the benefit of hindsight, the procurement strategy adopted by the energy supplier should have reflected a broader understanding of the risks associated with a former power station. The risks associated with meeting an immoveable time target in supplying power requires considerable expertise to price the work with sufficient accuracy and contingency. Given the recession at the time of tendering, winning any new work was seen as the position to be in. Risk and reward go hand in hand and the truely incentive based contract tries to ensure that all risks are fully appreciated and understood by the tenderers. From our perspective maybe these risks were not fully identified in this case.

CASE STUDY 12

Sector: utilities (NI)	**Project type:** waste water treatment works Stage II – secondary treatment
Project description: the construction of the second stage of the new treatment works covering extensive civil engineering groundworks.	

Objective of the scheme: to encourage innovation to beat time and budget constraints.

Background information: Stage I had run into problems using a conventional approach. This was to be avoided, if possible, on this next stage. To introduce some flexibility, the client encouraged the submission of alternative designs which would permit a fixed price alternative. In Northern Ireland the water utility service remains a nationalised industry within the DETR (NI).

Incentivisation category: responsive	**Overall project value:** £24 million
Overall outcome (success/partial success/failure) Success.	**Measurable benefits (time/cost/quality)** **Expected** – within budget and programme. **Achieved** – fixed price and with no variations. Completed on schedule.
Form of contract: M&E – General Conditions of Contract – Model Form A Civils – ICE 6th Edition.	

Description of scheme documentation: alternative tender created fixed price contract.

By separate agreement: none.

Operation of the scheme – administartive resource needed: greater focus on detailed design and the resolution of problems. Good co-operation meant the use of minimal resources to make payments at agreed milestones. Focus on improving on previous stage.

Evidence: views expressed in interviews/documents

Client/consultant: the failure of Stage I to live up to expectations had left its mark with the client team. Soured site relationships, heavy dependence upon documentation leading to a dispute with arbitration was not the ideal way to launch the second stage. Still using the ICE 6th edition and the M&E Form A the opportunity was taken to seek a fresh start. The results of this decision are abundantly clear on Stage II. There is trust between the parties, more open discussion takes place and the communication channels are open right up to board level.

The client believes this approach will set a precedent and encourage others to follow suit. The selection of the contractor was based upon the quality of their presentation, the clarity of their proposals and the expression of a no claims philosophy.

The fixing of the price allowed a terminal bonus to be set up upon achieving the revised time targets. The optimistic programme was achieved and the achievement of the win/win situation was a motivating factor in the project team's mind.

Contractor: the contractor was very confident of its own capabilities. This was manifested in the submission of an alternative design based on techniques developed by the contractor on previous contracts. This confidence was reflected in the submission of the lowest tender with the additional benefit to the client of being a fixed price.

There was a clear indication of good cooperation between the parties, since they did not wish to see a repetition of the Stage I problems.

How was the incentive scheme developed and handled: working at board level, the principles of the deal were discussed and the attraction of the fixed price to the client was a significant incentive to recommend acceptance. Within the traditional contract form, the idea of operating a partnership arrangement and an incentivised approach to implementing the works, was seen as innovative by all parties. This created the right environment for the scheme to develop and achieve the desired result.

Management of the incentive scheme: with little difficulty, the payment profile on achievement of certain milestones is a very simple payment mechanism to operate. Performance is crucial and the team's resources can properly focus on the job in hand, namely the construction of the works.

What improvements could be made on future projects of a similar nature: the form of contract is to be reviewed by the client, with the possibility of using the NEC on the next stage. The below ground risks should be separated from those above ground where more certainty exists. There needs to be a conscious effort to streamline the Government's procurement process since the annualised cash budgets curtail the number of opportunities to explore greater value for money.

Any additional information for future reference: the use of extended arm contracting is seen as an opportunity to be explored by the Government departments responsible for procurement.

Research contractor's comments: on multi-stage projects, understanding the reasons for the failure of previous stages contributes significantly to the fresh approach adopted by the parties. The balance of an assured design with the fixing of the price developed a win/win situation which fostered the right relationships from the beginning. Recognising that problems need to be jointly solved aids communication and activates the partnering approach.

CASE STUDY 13

Sector: civils infrastructure	**Project type:** road scheme – underpass and feeder roads
Project description: the construction of heavy civils infrastructure in and around residential properties.	

Objective of the incentive scheme: to overcome the accumulated delay and provide for a bonus for early completion so as to provide the vital access into a high priority development area.

Background information: this area was subject to restrictions on noise and vibration levels and all import and export of materials was to be by river to avoid disruption to the local road network. No incentives were included within the original contract and LADs were up to £40 000 per day. Problems had arisen during the first six months of the original contract due to design revisions and environmental constraints. These combined to give a projected delay of 1 year to the works.

Incentivisation category: reactive	**Overall project value:** £255 million
Overall outcome (success/partial success/failure) Success.	**Measurable benefits (time/cost/quality)** **Expected** – to beat the time and cost targets **Achieved** – completed five months early and on budget.
Form of Contract: GC Works 1 with Re-measurable Bills of Quantities.	

Description of scheme documentation: the ICE Standard method of measurement was used for the civil works with a remeasurement process and a Cl 12 arrangement included. Essentially it created a lump sum price in the final analysis. The contractor's payment for overcoming delay was partly performance related – with a bonus for early completion. Value engineering was introduced with an agreed share of the savings in cost.

By separate agreement: the incentivisation was written into a separate agreement.

Operation of the scheme – administrative resource needed: no additional resource was needed over and above that required to operate the works contract. The value engineering process was designed to operate on the basis of an offer rather than a cost and expense claim.

Evidence: views expressed in interviews/documents

Client/consultant: initially, the client and the resident engineering team were wary of possible contractual exploitation by the contractor. This level of mistrust only added to the tensions in managing the contract. The traditional approach to design and build led to positions being defended and caused delay to the programme. The recognition that changes in the management style had to occur, led to the acceleration programme and the introduction of key specialist project managers acting for the client. This facilitated the new agreement. The consultant had previously carried out the design and supervision roles. The new approach required them to focus on the design only.

Contractor: the contractor faced major construction and environmental risks. From the outset, the relationships between client and contractor staff was strained. The contract terms were rigidly applied by the consultant which caused tensions to develop.

The revised approach introducing a lump sum price contributed to a radical re-think of the construction management team. This improved the working relationships and developed a more trusting environment. It was a significant contribution to the successful outcome of the project.

How was the incentive scheme developed and handled: having quantified the delay, agreement was reached at Board level on measures to deal with this delay and to settle claims that had arisen. A joint approach was adopted to accelerate completion with a bonus payable for overcoming delay and , if possible, completing earlier than the original contract date.

With the change in circumstances, the speed of resolution of problems was imperative if the new timescales were to be achieved. Periodic reviews were introduced to check progress on key issues.

Management of the incentive scheme: the management arrangements were revised to provide a quarterly review of progress and problems by a panel empowered to make decisions on measures to be taken. With the acceleration agreement in place, the Value Engineering was put in place to assist the acceleration.

Following successful implementation of these measures, further changes were made to ensure that the necessary power to make decisions/awards was vested in the client's project manager.

What improvements could be made on future projects of a similar nature: in practice, on large projects involving innovative techniques the assessment of risk is critical if such risks are to be shared. Until the extent of the problems are determined, it is likely that some period of risk review may be necessary before assigning the risks. In this case, once the problems of the project had been uncovered, an incentive scheme was introduced. Coupled with this action, amendments were made to the contract to allow it to operate successfully.

Any additional information for future reference: this case raises the importance of risk assessment and the process of incentivisation. Incentivising performance can only be done if risks have been properly assigned or that there is a process in place to handle unforseen risks.

Research contractor's comments: looking at the action taken overall, there are some clear messages which stand out:

- it is not always obvious that an incentive scheme will be required to enable the project to succeed

- the project team need to be collectively responsible for monitoring problems as they arise and there should be mechanisms in place to allow these to be brought to the surface, rather than being buried and hope they go away

- the client and the contractor should encourage openness and trust which deve-lops a more meaningful working environment. this often leads to improvement in performance and a better quality job.

CASE STUDY 14

Sector: light processing	**Project type:** new semi-conductor manufacturing facility
Project description: to construct a new manufacturing plant in the north of England.	

Objective of the incentive scheme: to ensure that the target date for the opening of the new plant was met.

Background information: the demand for microchips is increasing. New designs need to be manufactured quickly if suppliers are to capture the market. To meet this expected growth in demand, any manufacturer must be able to guarantee the commencement of production to match the market needs so as to secure the supply contracts. With competition from the Far East and Europe, these plants must come on stream on time to meet customer demands. Time is of the essence. With such a volatile market, it is imperative that there is a close alignment of objectives between the client and the contractor to rise to the challenge. With the potential for the client to apply LADs at £1 million per day, the margin for error by the contractor is small.

Incentivisation category: pre-planned	**Overall project value:** £170 million
Overall outcome (success/partial success/failure) Success.	**Measurable benefits (time/cost/quality)** **Expected** – vital to meet or beat the target date. To keep within the target cost (52 week contract). **Achieved** – opened on time and within budget.
Form of contract: JCT Design and Build (heavily modified).	

Description of scheme documentation: the only document which existed at the start of the contract was a letter of intent from the client stating that the contractor was awarded the contract with a guaranteed maximum price. Since the contractor had recently constructed a similar plant in Germany, their track record weighed heavily in their favour. Documentation was kept to a minimum since it was essential to commence work to meet the tight time deadlines. Using an open book approach on the procurement of packages of work, the monitoring of costs and the payment of fees was based on milestone achievements.

The overall cost was built up by a fixed element for design, management costs and fee together with the specialist contractor costs. These were paid in accordance with the agreed milestone payment schedule.

By separate agreement: none.

Operation of the scheme – administrative resource needed: minimal resources were applied since the client and contractor were monitoring performance only on a milestone basis. Design changes were agreed and costed throughout the project.

Evidence: views expressed in interviews/documents

Client/consultant: the client needed the manufacturing capacity to enable them to beat other major competitors in world markets. They were clearly focussed on the selection criteria for determining the most appropriate contractor.

Within the private sector, the accountability is towards the shareholders and if the business opportunity was missed, then market share may not be recovered. The aim was to select a contractor which understood the market and was technically able to deal with the complexities of a high-tech factory structure and the installation of specialised equipment.

Contractor: with such a highly technical processing plant, there were significant risks in creating new manufacturing facilities to develop new products. These risks were minimised to the contractor by the use of a design methodology which enabled the client's team to be an integral part of the development process.

Working on a budget ceiling was intended to act as a control mechanism and was not intended to transfer risk to the contractor if significant changes had to be made to meet the time deadline. Throughout the project programme the emphasis was on meeting the technical specification to match the client's demands.

How was the incentive scheme developed and handled: the scheme was based on shared incentives within the design and development process. The guaranteed maximum price enabled the contractor to fix their fee for managing the construction of the plant which allowed them to determine the resource base more accurately. The changes in design were thus handled within the target cost with the risks shared between the parties. This encouraged a value engineering approach where savings in one area would provide the means to allow increases in other areas. This focussed the management of the project with the key personnel since they were empowered to make decisions. This disciplined approach was the key to minimising disruption to the building programme.

Management of the incentive scheme: with the low level of paper communication, the incentive scheme was managed on the basis of performance which relied on a system known as "Records of Discussion". These were short minutes of meetings/ conversations typed on orange paper. Thus if an orange paper was received, it was important for the recipient to read it urgently. This type of communication covered the intended design change and the cost implication.

What improvements could be made on future projects of a similar nature: none identified, since both the client and contractor were very satisfied with the outcome.

Any additional information for future reference: this approach is very similar in style to partnering, where the key personnel are an essential part of the deal, since they have either worked together before or they have an impressive track record within the industry.

Research contractor's comments:

The private sector has a greater degree of freedom to enter into such a contract in this way, since it is driven by market and commercial considerations. However the principles by which the project was conceived and managed demonstrates that where the design and construction risks are at least shared then there is a greater propensity to develop mutual trust and focus on the central issues. In this case the achievement of the opening date was crucial and all efforts were focussed towards this common objective.

CASE STUDY 15

Sector: utilities	**Project type:** wastewater treatment works
Project description: the construction of a new sewage treatment works on a greenfield site to meet the demands of European legislation under the Cleanwater Directive.	

Objective of the incentive scheme: to ensure that the operational date was met so as to comply with the EC Directive.

Background information: the Water Companies are subject to increasing EC demands to provide improved treatment of effluent prior to disposal at sea or into the local water environment. The EC legislation placed the responsibility for meeting the discharge requirements with the Water Companies and they in turn needed new or improved facilities to handle these requirements. This gave rise to a rapid demand for facilities within the sector which were needed within very tight time deadlines. Time was of the essence to meet the EC demands. The approach used by some water companies was extended arm contracting, whereby they could mobilise projects relatively quickly. The development of output based specifications allowed the client to use the design and build concept more extensively and adapt them to include incentives as appropriate.

This project was selected as one of two case studies (the other being Case study 16) from the same water company to determine whether any striking differences arose from applying a similar contract strategy on different sites.

Incentivisation category: pre-planned	**Overall project value:** £12 million
Overall outcome (success/partial success/failure) Partial success.	**Measurable benefits (time/cost/quality):** **Expected:** to complete the works on time and within budget (21 months contract). **Achieved:** whilst operational consent was granted by the due date, the overall programme overran but was completed within budget.
Form of contract: IChemE *"Green book"* for the construction works ACE Agreement for the Design.	

Description of scheme documentation: based on a target cost developed during an open book negotiation. The design and build approach enabled a fast track programme to be followed. In addition to the application of LADs as negative incentives a gain share arrangement was introduced which applied when targets were beaten.

By separate agreement: an additional umbrella arrangement was documented to establish the partnering approach. This was a separate, non-legal, non-binding document to crystallise the working arrangements between client project manager, designer and contractor.

Operation of the scheme – administrative resource needed: a close working arrangement existed as a result of the umbrella arrangement, with a common site office for the core team. Paper communications were thus minimised and recording decisions were agreed.

The culture within the contractor's organisation was such as to closely align their objectives with the client. The contract document was not seen as the driving force for the project.

Evidence: views expressed in interviews/documents

Client/consultant: the preparation of a detailed scope of works using the principles of standardised design assisted the determination of the target cost. Where the scope was not sufficiently detailed, then this placed more risk with the contractor.

The development of design followed a two stage process:

- Stage I – following the feasibility study an outline design was prepared. At this point the contractor was invited to comment and be part of a value engineering exercise. The scope was agreed to allow the target cost to be calculated
- Stage II – detailed design was carried out create the construction drawings and plant specifications.

The achievement of obtaining the consent to operate the plant was a major milestone both in terms of bringing the plant on stream and for payment to the contractor. The subsequent delay in completing the work within the contracted time is to be regretted, but the prime objective had been fulfilled. The statement of partial success needs to be qualified in this respect.

Contractor: the contractor was well geared to build the works with no inhibitions as to the working arrangements. Within the target cost for this greenfield site, there were arguments for and against the approach adopted. Generally the contractor favoured the approach taken.

With a buoyant market, prices were competitive and relied upon the selection of the most appropriate partners to build the scheme. Having a track record of good working relationships added to the choice of contractor. With design and construction the contractor is looking to safeguard profitability and with the consequent lowered risks the contractor can operate at lower margin.

All profitable projects are good projects when the client's objectives are met.

How was the incentive scheme developed and handled: the client initiated the scheme based on previous experience. However this was not difficult to set up but the success of the scheme relied on more detailed input to the output specification to remove as much risk as possible in the procurement of plant and equipment. The lower the risk, the greater the focus on performance, and hence meeting the objectives.

Management of the incentive scheme: this was not a significant issue since the close working relationships enabled the programme to be closely monitored and the achievements of milestones used as part of the payment mechanism. Overall focus on the resolution of problems and the alignment of objectives meant that the project issues were dealt with as speedily as possible and therefore there was no opportunity for confrontation to arise.

What improvements could be made on future projects of a similar nature: as with Case study 16, improvements were identified for future schemes.

1. Determine the right target estimate so as to produce a value for money solution for all parties.

2. Greater definition of the scope will reduce the likelihood of poor design and improve the specification of how the plant is to be controlled, what operational detail is needed and what type of pumps are required.

In this way the contractor can form alliances with manufacturers within framework agreements. In addition this also allows for optimisation of the equipment specifications, which is a critical feature of this type of facility.

Any additional information for future reference: none.

Research contractor's comments: this project demonstrates the evolving procurement methods adopted by this client to overcome disruption and delay to projects. It draws together some of the basic arguments to introduce incentivisation into the contract, namely: clear objectives, detailed output specifications, agreed target cost, good working relationships and a methodology based on a partnering approach.

The disappointment was that the overall contract works were not completed within the set contract period. On balance this increased the contractor's overheads and caused the client to maintain a project team for longer than anticipated which in turn meant additional cost. However, it has to be re-stated that the prime objective of gaining operational consent on time was achieved.

CASE STUDY 16

Sector: utilities	**Project type:** wastewater treatment works
Project description: the construction of a new sewage treatment works on a greenfield site to meet the demands of European legislation under the Cleanwater Directive.	

Objective of the incentive scheme: to ensure that the operational date was met so as to comply with the EC Directive.

Background information: the Water Companies are subject to increasing EC demands to provide improved treatment of effluent prior to disposal at sea or into the local water environment. The EC legislation placed the responsibility for meeting the discharge requirements with the Water Companies and they in turn needed new or improved facilities to handle these requirements. This gave rise to a rapid demand for facilities within the sector which were needed within very tight time deadlines. Time was of the essence to meet the EC demands.

The contract strategy adopted by some Water Companies was to use extended arm contracting, whereby they could mobilise projects relatively quickly. This Water Company had developed a different approach using a standardised detailed output based specification which allowed them to shorten the time to mobilise each project on a design and build basis. The incentivisation schemes also followed a similar format, based on a target cost with incentives for early completion with a shared savings scheme. Since the majority of the capital projects had to be completed by rigidly enforced dates, time was of the essence. This project was selected as one of two case studies (the other being Case study 15) from the same water company to determine whether any striking differences arose from applying a similar contract strategy on different sites.

Incentivisation category: pre-planned	**Overall project value:** £24.5 million
Overall outcome (success/partial success/failure) Partial success.	**Measurable benefits (time/cost/quality)** **Expected:** completion within 104-week contract period and within budget. **Achieved:** whilst operational consent was granted by the due date, the overall programme overran by 18months even allowing for a nine-month extension due to unforeseen circumstances. The final cost was within budget.
Form of contract: IChemE *"Green book"* for the design and construction works plus modifications (Clause 12 built in). An extensive 14 volume document based on using a standardised company approach.	

Description of scheme documentation: a design and construct contract with a target cost negotiated on an open book basis. The design and build approach enabled a fast track programme to be followed since the risks were largely known. In addition to the application of LADs as negative incentives a gain share arrangement was introduced which applied when cost targets were beaten.

By separate agreement: none.

Operation of the scheme – administrative resource needed: a close working arrangement existed between the client and the contractor with a common site office for the core team. As such, paper communications were minimised, recording decisions agreed. The contract document was not seen as the driving force for the project since the working arrangements were in good shape.

Evidence: views expressed in interviews/documents

Client/consultant: based on the client's concept design, a detailed scope of works was prepared using the principles of standardised design elements. This greatly assisted in the determination of the target cost.

The contract was tendered on a concept design with a Bill of quantities. Alternative designs were accepted. From the six tenders received, the prices were a dominant feature. The initial analysis showed that the lowest tender had an acceptable design. Following a six-month design assessment, the final two tenderers were asked to submit best and final offers to confirm the final target cost.

All the risks were largely known, as the site was adjacent to the existing outfall. However, a serious problem arose when 138 large concrete tank traps were discovered in random locations. All known references had not indicated their presence. This required extensive excavation which extended the contract period by nine months. The additional costs were handled under the Clause 12 (unforeseen circumstances) arrangements.

The shared savings scheme was instigated to encourage the contractor throughout the construction, to consider any savings which could be achieved. The split was 50:50.

Contractor: this was a complex construction with a subterranean treatment works with the control gear located above ground. In a planning sensitive area on the seafront, the completed building resembled a Martello Tower which is a Napoleonic defence structure. The odour control system ensured the acceptability of operating the plant on the particular site and the gas detection system was the latest high-technology.

Early delays caused by the tank traps added to the pressure on time to obtain operational consent by the set date. This was achieved since an additional 2 months float had been included to cover any potential delay.

Overall the quality of workmanship was good and the costs were contained within the target adjusted for the shared overrun period. The M&E works were completed to time, cost and quality. From the contractor's viewpoint the civils works were not completed on time and there were cost overruns. The quality was good with the superstructure receiving good publicity. The optimisation of the control equipment is a significant task during the commissioning phase.

Overall the contract programme submitted at the tender stage was too optimistic. With the benefit of hindsight a 2.5–3 year programme now seems more realistic.

How was the incentive scheme developed and handled: the client initiated the scheme based on previous experience. However this was not difficult to set up but the success of the scheme relied on more detailed input to the output specification to remove as much risk as possible in the procurement of plant and equipment. The lower the risk, the greater the focus on performance, and hence meeting the objectives.

Management of the incentive scheme: this was not a significant issue since the close working relationships enabled the programme to be closely monitored and the achievements of milestones used as part of the payment mechanism. Overall focus on the resolution of problems and the alignment of objectives meant that the project issues were dealt with as speedily as possible and therefore there was no opportunity for confrontation to arise.

What improvements could be made on future projects of a similar nature: as with Case study 15, improvements were identified for future schemes.

1. Determine the right target estimate so as to produce a value for money solution for all parties.

2. Greater definition of the scope will reduce the likelihood of poor design and improve the specification of how the plant is to be controlled, what operational detail is needed and what type of pumps are required.

In this way the contractor can form alliances with manufacturers within framework agreements. In addition this also allows for optimisation of the equipment specifications, which is a critical feature of this type of facility.

The argument for a time based bonus scheme should be considered since in some cases the spend to save principle can dramatically improve the chances of meeting the time deadline. Using LADs as a negative incentive is unlikely to encourage the necessary changes in productivity levels without incurring additional costs.

Any additional information for future reference: none.

Research contractor's comments: this project demonstrates the evolving procure-ment methods developed by this client to overcome disruption and delay to projects. It draws together some of the basic arguments for introducing incentivisation into the contract, namely: clear objectives, detailed output specifications, agreed target cost, good working relationships and a methodology based on a partnering approach.

The disappointment was that the overall contract works were not completed within the set contract period. On balance this increased the contractor's overheads and caused the client to maintain a project team for longer than anticipated which in turn mean additional cost. However, it has to be re-stated that the prime objective of gaining operational consent on time was achieved.

For this client, the benefits of using the IChemE *"Green book"* form of contract with a shared incentive scheme demonstrates their commitment to the partnering approach. Within this industry sector, there are politically sensitive issues which would have European ramifications if performance did not meet the required Directives.

CASE STUDY 17

Sector: civils infrastructure – private sector	Project type: fast ferry terminals
Project description: the construction of terminals for the new generation of fast ferries.	

Objective of the incentive scheme: to meet or beat time targets to ensure that the ports were able to meet increased passenger demands.

Background information: since the early 1990s, fast roll-on/roll-off (RoRo) vessels have been introduced on a number of key ferry routes in the English Channel, Irish Sea, North Sea and Baltic. The new vessels have changed the nature of the RoRo market and have been the catalyst to a number of significant port development projects. Terminal facilities must be planned as a whole or individual berths may not realise their potential. A key feature of many of the fast ferry berth projects under-taken to date has been that development of berth concepts has been required to proceed in parallel with construction of the new ships. This challenge demands a very flexible approach to scheme planning, design and construction in order that necessary changes in concept can be accommodated without causing delays or disproportionate cost increases.

Incentivisation: pre-planned	Overall project value: £45 million
Overall outcome (success/partial success/failure) Success (an outstanding success which encapsulated the combined efforts of all parties).	**Benefits (time/cost/quality)** **Expected:** 16-month programme and within budget. **Achieved:** Completed three months early and within budget. No claims throughout the project. Final out-turn just below £40 million.
Form of contract: ICE 6th edition Design & Build with modifications.	

Description of scheme documentation: Clauses 40 and 41 were modified and the client's terms and conditions were incorporated. In adopting a fast-track approach, the client appointed the consultant to act on their behalf on all project matters. This entailed the management of some 18 specialist contractors. To highlight the time pressures, the client introduced LADs of £10 000 per day on all contractors. In effect there were 18 design and build contractors working to the same terms and conditions.

By separate agreement: none.

Operation of the scheme – administration resource needed: while the traditional re-measurement approach was used for the lump sum items and a schedule of rates used for the variable elements to arrive at the monthly valuation, the project team's efforts were directed towards problem resolution to maintain adherence to the programme. For design issues, problems were referred back to the consultant's office. Construction problems were dealt with on site. Instructions were issued orally and acted upon, with follow-up memoranda to confirm instructions. The monthly valuation covered the remeasure and cost of variations. Costs were controlled using an updating approach to the estimate incorporating a change control procedure.

Evidence: views expressed in interviews/documents

Client/consultant: the increasing capacity provided by new ferry designs put pressure on the port facilities. The largest fast ferry is capable of carrying up to 350 cars or up to 50 trucks and 100 cars. These ferries incorporate several novel features designed to reduce the turnaround time at the berth to the absolute minimum. In addition, the operator required standards of accommodation and level of service on board and in the port to be improved to a level similar to that which passengers would expect in air travel. The construction of facilities for fast ferry operations, therefore has to be undertaken to demanding programmes subject to restrictions on working methods resulting from the need to avoid disruption to operations during construction. In this case, the challenge is accentuated by the commercial imperatives.

The experience of this project is that technically complex projects can be successfully completed to the tightest programmes provided designs are practical and address the underlying project objectives and that thorough investigations and preparatory works are undertaken prior to commencing work.

With a performance specification based on outputs, the price for each package of work was not the central issue but the capability of meeting the exacting time deadlines was paramount.

To ensure the scheme's success, the composition and workings of the project team was all important.

Contractor: each of the specialist contractors formed part of the overall team with the client and consultant working closely alongside. The employer, port operational personnel, designer and contractors worked together to overcome problems which arose during construction.

The combination of good design and the constructive approach to problem solving was the key to the successful delivery of this project to time and within budget and without contractual conflict or disruption to ongoing port operations. Where the client was not sure of their precise requirements, then each contractor was allowed to use the best method of construction within the permitted design parameters. The emphasis during the tendering exercise was to secure the commission based on a sound track record and a price which reflected the buildability of the solution.

The tender prices were extremely competitive, demonstrating that the risks had been fully identified. Whilst risk was transferred, the resolution of problems was the key to success.

How was the incentive scheme developed and handled: the initiative to introduce an incentivised contract was focussed by the client and seized upon by the contractors. The consultant was involved in ensuring that the parameters for the scheme were properly explained to minimise uncertainty. Risk analysis, value management and contract management were all applied by the client to demonstrate the robustness of the approach. With potential repeat work the contractors were keen to be as open as possible during the tender negotiations. Programme control was exercised by formal meetings, site discussions and focussed target meetings. Potential conflict was removed by the development of trust in each of the parties capability.

Management of the incentive scheme: involvement at director level ensured that a high level focus was maintained on both the client's and contractor's teams. Time was rigidly controlled which set the parameters for the design and construction process. There was constant pressure to look for ease of construction and improved prog-ramme time for construction. This alignment of objectives was further enhanced by the safeguarding of the principle of the contractor's cost + overhead + profit within the overall picture.

What improvements could be made on future projects of a similar nature: to improve on the communication and resolution of problems, the use of more senior project managers would have speeded up the process. The high level of adminis-trative support is inevitable on this level of interation on a D& B project. Whilst this was needed it was kept subservient to the prime objective of time. A more stream-lined change control procedure may have speeded up the design approval process.

Any additional information for future reference: this project was part of an overall series of projects designed to meet the needs of the fast-ferry market. With extremely tight time deadlines, the incentive is to reward delivery once designs have been approved. Where no one specialist contractor dominates the team, the interpersonal issues are of equal importance in ensuring that the various parties can work together and achieve the prime objective. During the tender evaluation process, this capability should be clearly demonstrated, both by individuals achievements and clear testimonies from previous clients. Any doubts should be thoroughly investigated prior to contract award.

Research contractor's comments: there are significant construction risks to be managed on design and build projects of this nature. In- built flexibility adopted by both client and contractors, meant that each recognised the difficulties likely to be faced by other parties. This recognition manifested itself in the development of a strong bond between the personnel working on the project. Ownership of problems was shared and the real incentive was to meet or beat the time targets. This approach was subsequently repeated on other terminal site projects. The basic ingredients can be identified.

1. A clear modus operandi within which issues were to be resolved.
2. Key personnel used with clearly focussed objectives.
3. Maximise team performance to achieve a win/win situation.
4. Alignment of objectives to achieve best value solutions.

The marine environment is very harsh with a design life on the main structural elements of the order of 25 years. The consumable elements(eg fendering) are likely to last only 12 years. Therefore the design demands must take into account the ease of maintenance of critical items. Like any train or airline operator, the non-availability of proper berthing/stabling facilities causes immediate delay which is directly felt by the passengers. The completion of new facilities therefore brings direct benefits to the operator and enhances the credibility of the construction industry.

CASE STUDY 18

Sector: buildings	**Project type:** leisure parks
Project description: to construct new leisure parks on various sites throughout the UK.	

Objective of the scheme: to ensure that the parks are opened on time to capture customer demand.

Background information: there are several key features in the development of such parks which create the environment for the type of contract used in the construction of the buildings and the incentives needed to ensure that programmed dates are met. The critical issue is one of securing planning consent. This generally takes a considerable amount of time and is heavily dependent upon public consultation and the support of the local authority. Once planning consent is granted, then the project has to be mobilised very quickly. This time pressure is felt all down the line from the client to the sub-contractors. The general accommodation can include cinemas, restaurants, bars, bowling alleys, bingo halls and nightclubs. These are all special buildings and require expertise in installing equipment for the different purposes.

Incentivisation: pre-planned.	**Overall project value:** projects vary between £5 million to £40 million.
Overall outcome (success/partial success/failure) Success.	**Measurable benefits (time/cost/quality)** **Expected** – generally 12–15 months. **Achieved** – on time and within budget.
Form of contract: JCT 80 – Design & Build.	

Description of scheme documentation: the traditional form of contract included LADs which were to be applied in the event of a programme overrun. Whilst the lump sum price was fixed, in reality, this was not the case since the client initiated a considerable amount of value engineering to drive down costs. The price negotiation included a significant percentage for contract preliminaries and profit, which guaranteed a 5 per cent margin. However by accepting the negotiated pricing route this was not always achievable.

By separate agreement: roles and responsibilities are covered by a separate agreement to identify key activities and the approach to be taken.

Operation of the scheme – administrative resource needed: the independent monitoring of costs by the quantity surveyors was a resource used by the client to keep the pressure on the costs. Following value engineering exercises, if the costs went up then these were negotiated, if the costs were less then the savings were shared 50:50.

Evidence: views expressed in interviews/documents

Client/consultant: there have been several projects of this nature completed so far, with two per annum planned for the near future, with some £0.25 billion investment. This gives the client extensive buying power in the private sector market. The development of close working partners is part of a contract strategy, which has to be mobilised once planning consent is granted. There is a need to insulate the market when the client needs to move quickly and, by negotiating the price, it saves time and effort. It benefits the overall programme and retains the specialist building skills for this type of development.

The complexity of the buildings requires a broad range of skills. Not many contractors have this track record and hence the need to maintain good partnering arrangements. In some sense this procurement strategy is very similar to the approach operated by Marks and Spencers whereby price, quality and delivery are all fixed with little room for the suppliers to reflect significant cost changes to their raw materials.

Contractor: with this partnering arrangement, the contractor is pricing the risk. With LADs and the approach to value engineering, there is little opportunity to make any significant margins on this work. This is not conducive to aligned project objectives. However the expected volume of business over the foreseeable future means that a large amount of work can be guaranteed, if the price is right. The client demands means that a very tight programme was agreed to meet the opening date. This creates excessive pressure on all contractor's staff and can have a negative effect. The threat to changing the partnership arrangements means that the contractor generally agrees to client demands.

How was the incentive scheme developed and handled: largely by negotiation within the framework of previous projects. The client is in the driving seat with such a volume of business to trade with in the industry. The pressure on time is highlighted by the use of LADs for overrun. Cost is not as critical, although important. Acceptance by the contractor to the pricing terms within the context of the private sector relates to the harsh commercial realities in contrast to the overall value for money approach within the public sector.

Management of the incentive scheme: the independent consultants and the quantity surveyors acted for the client. This resource is not any larger for this type of project than is normal. Regular progress meetings and the pressure on time meant that all resources were focussed on this target. With the opening date as the critical milestone, most other considerations were of lesser importance.

What improvements could be made on future projects of a similar nature: if more investors come into the market place, then other opportunities for the contractors may become apparent. If this situation arises, then there is likely to be a more balanced negotiation as to terms and conditions. The manner in which the private sector operates on this type of project does not directly relate to typical situations within the public sector.

Any additional information for future reference: none.

Research contractor's comments: this is an approach adopted by the private sector when the long-term investment profile is massive. With the concentration of specialist skills in only a few contractors, it becomes a niche market. The development of partnering arrangements is seen as the means of keeping the skills intact and the balance of the negotiation seems to reside with the client. For how long this situation can continue probably will be determined by the outcome of the projects. If projects start to overrun in terms of time and cost it is likely that the negotiations will become more balanced. Commercially, no contractor could sustain for any length of time continued losses on projects even though there was continuity of work. Perhaps it is this tightrope which is missing in some public sector area to increase value for money. On balance this is not the way to go, since a truly incentivised contract will afford the contractor every opportunity to perform and make a profit.

CASE STUDY 19

Sector: transportation – private sector	**Project type:** highways long-term maintenance scheme
Project description: the construction of a new bypass and other improvements together with the management and maintenance of a stretch of trunk road.	

Objective of the incentive scheme: over a thirty year period, to encourage innovation and improvement in the operation of the highway, with particular emphasis on improved road safety.

Background information: as part of the Government's Private Finance Initiative, responsibility for managing and maintaining the trunk road was transferred to a special purpose company (SPV), a consortium of private companies. Using its skills, technical knowledge and expertise this company is responsible for ensuring that the road and its environment are kept in the very best condition for the next 30 years. It is important to realise that the road still belongs to the Government – it has not been sold to the private sector. The Highways Agency will still be the responsible authority and will ensure that the SPV carry out their maintenance obligations to the very highest standard. All major decisions on how public money will be spent on develo-ping and improving the road will continue to be made by the Highways Agency.

This contract was included in the list of case studies to gain an insight into how risk and reward are handled over a long contract period. Whilst contractual obligations have to be fulfilled, the risks have to be adequately assessed if the scheme is to be affordable when compared with traditionally let maintenance contracts.

Incentivisation category: pre-planned	**Overall project value:** capital works £9.4 million. Maintenance payments based on shadow tolls
Overall outcome (success/partial success/failure) Success.	**Measurable benefits (time/cost/quality)** **Expected** – new works, within budget and to be operational by the due date. **Achieved** – completed two months early and within budget.
Form of contract: unique form of contract to cover a 30 year concession agreement.	

Description of scheme documentation: payments throughout the contract are based primarily on traffic flows. The client will pay the different "shadow tolls" to the contractor according to the level of traffic. Payments can be adjusted to reflect performance, particularly the number and timing of lane closures for maintenance or other purposes. The contract allows bonus payments to be made when the contractor comes forward with agreed investment which can be demonstrated to have improved highway safety.

By separate agreement: specific terms.

Operation of the scheme – administrative resource needed: with a long contract period, good record keeping is paramount. With information technology this task is assisted particularly in the capture of highway data and the key issue is to monitor performance. Whilst the contractor has the day-to-day responsibility for running the contract, the client has to ensure that the service is meeting the required quality. The close working relationship of the team keeps the amount of resource to a minimum.

Evidence: views expressed in interviews/documents

Client/consultant: the client's principal role is to monitor the performance of the contractor, taking appropriate action as necessary. Quality assurance and safety standards are set by the client and must be followed throughout the contract period. The client has the right of audit and inspection and will monitor the road regularly.

The preparation of the contract required considerable effort since there was no convenient precedent. The contract had to address the new operating environment and achieve the necessary balance between of rights and reponsibilities between the parties.

There had been some criticism of the the lengthy tendering process, with complaints about the expense and time involved. It is felt that as parties become more familiar with the DBFO roads procedure, the process will speed up and costs will reduce. To measure the quality of response by the contractor a series of performance indicators have been agreed.

Contractor: all those engaged in the negotiations of DBFO road projects, as well as the banks considering lending money to bidding consortia, are all becoming more versed in the particular issues which are raised on road schemes. The effort required from conceptual planning to award of contract will accelerate with an increased understanding of each other's objectives. With such a long contract it is imperative to ensure that if problems do arise that there are sufficient mechanisms to resolve them.

The construction of the bypass and associated structures has gone well and under normal circumstances would have been the main focus of attention. When included within the whole maintenance and operation requirement it becomes a part of a larger contract. However good performance sets the right approach for the remainder of the concession.

For day-to-day tasks of road naintenance we have entered into an agreement with a consortium of three local authorities in the area. In addition they also provide a Winter Maintenance Service. Both parties are committed to working together to find ways to improve delivery of these essential services to the public.

How was the incentive scheme developed and handled: there was a relatively long lead time to the project. It took some 20 months from the decision to proceed to the signing of contracts. Whilst the basic requirements were identified early on the detailed arrangements and safeguards had to be precisely written in a new form. Substantial legal advice was required to translate the detailed engineering requirement into an appropriate tender format. The financial model is complex, taking into consideration the whole life costs.

Management of the incentive scheme: payments can be adjusted to reflect performance, particularly the number and timing of lane closures for maintenance or other purposes. Any proposals to improve safety have to be reviewed to establish current benchmarks, since if approved then the payment of any bonus will need to reflect actual achievement from the previous situation.

To indicate the variety of performance indicators, the following list covers the key issues:

- maintaining roads and footways
- maintaining street lighting
- emergency call outs
- winter maintenance services
- management of roadworks
- requests for services

What improvements could be made on future projects of a similar nature: the high cost of bidding for schemes is the main stumbling block for the industry. Costs could be reduced by standard contract documents being produced and the process could be speeded up by producing – and sticking to – a tighter timetable. In future schemes, the Government also proposes to identify its preferred bidder at an earlier stage.

Any additional information for future reference: all long-term risks may not yet be fully understood and who is best placed to manage them. Only time will tell.

Research contractor's comments: the pathfinder status of this project is likely to serve as a benchmark for the future. Whilst the contract period is very long, it is worthwhile noting the key issues which set DBFO projects in the context of the incentivisation of construction contracts.

1. The assessment of risk and who shall manage the risks is the bedrock of public/private partnerships.
2. Setting aside the risk issues, the payment mechanism has to balance risk and reward with incentives built as appropriate to gain improvement in standard performance.
3. The lengthy negotiations with the preferred bidder creates the right environment for the contract to succeed, since there has to be agreement on the alignment of objectives for the overall scheme to move forward.
4. What matters in the management of the contract is not the contract form but the clarity of the key issues and the responsibilities of all the parties involved.

CASE STUDY 20

Sector: buildings	**Project type:** multi-storey offices
Project description: to construct a town centre office block for a variety of prospective tenants.	

Objective of the incentive scheme: to complete the building within tight time parameters to ensure that the pre-let arrangements could be achieved and the building occupied as soon as possible.

Background information: the design and construction of high rise offices like any tall building has different types of risk at different stages of the construction. The client realised that to overcome any potential problems, there was the need for a partnering arrangement so that the risks could be properly allocated and managed. With the decision to adopt this approach, there was a considerable amount of setting up and agreement to be reached. To assist in this development, the client initiated a value management workshop to discuss the issues and reach a consensus viewpoint.

The site is located within a large city centre and is within a conservation area. Prior to demolition, it was occupied by a 1970s tower block which had come to the end of its useful life. Options to up-grade the site ranged from simple repainting to full development of the site.

Incentivisation: pre-planned	**Overall project value:** £10.7 million
Overall outcome (success/partial success/failure) Success.	**Benefits (time/cost/quality)** **Expected:** 72 week programme within guaranteed maximum price. **Achieved:** substantially completed on time and within budget.
Form of contract: JCT 80 – Two stage tendering process.	

Description of scheme documentation: in addition to the standard building contract document, special emphasis was placed on a separate partnering agreement setting out the way the various parties; client, designers, quantity surveyors, consultants, contractor and sub-contractors were to operate. This was based on a rapid dispute resolution technique coupled with regular monthly partnering meetings where a score sheet with ten objectives was completed by each party to monitor overall progress. In addition, four to five partnering champions were appointed to progress the issues as quickly as possible.

By separate agreement: a partnering charter stating the goals and objectives for the project was formalised and signed by all the key parties. This included a commitment to do everything in their power to achieve the goals and objectives.

Operation of the scheme – administration resource needed: the amount of resource needed to manage such a project was considerable. As part of their overall responsibilities, the partnering charter encompasses the key issues to incentivisation, namely, monitoring performance and resolving difficulties to everyone's satisfaction. The pro formas produced as part of the output from the value management workshop, showed a keen awareness of the critical success factors which had to be controlled if the project was to succeed.

Evidence: views expressed in interviews/documents

Client/consultant: based on a two stage tendering process, the concept design with a package of drawings was priced by tenderers. Following the selection process, including the evaluation of method statements, the successful tender went into the second stage. The tender was fully billed with detailed drawings and a guaranteed maximum price (GMP) was negotiated on an open book basis. The sub-contractors pricing achieved a best value solution by the selection of the best price. The client agreed the GMP with the client taking the risk of design.

Contractor: operating within a partnering arrangement can be less productive if it becomes just a talking shop. However the working arrangements have been satisfactory, since the contributions from all sides are focused on the prime objective and not just aimed at brow-beating the contractor. With the allocation of risk and the team building approach, there was every incentive to perform. The opportunity to receive a terminal bonus for meeting or beating the time deadline is a significant incentive, however the spend to achieve the deadline has to be monitored carefully to ensure costs are not excessive.

How was the incentive scheme developed and handled: through the normal tendering procedure, the LADs and the terminal bonus arrangement were defined and agreed by all parties. There was an incentive for all parties to receive if successful. For those on fees it amounted to 7 per cent of their fee in some cases.

Management of the incentive scheme: close monitoring of the construction programme allied with the efforts of the partnering arrangement have created an environment which is condusive to a proactive rather than a reactive stance being taken by all parties. The regular monthly meetings are used as a means of monitoring performance and using the dispute resolution process to reach agreement and to give direction.

What improvements could be made on future projects of a similar nature: bringing the contractor into the design and build process as early as possible contributes towards the team building process which creates the right working environment.

Any additional information for future reference: this approach has been successful and has aleady been used on another site with even better results.

Research contractor's comments: the amount of effort in setting out clear guidance on how goals and objectives are to be achieved seems to have been successful. Whilst LADs were included to make time to complete the building as the prime objective, terminal bonuses were added as a positive incentive for achieving that goal. No amount of documentation makes up for good working relationships, but in this case it has significantly contributed to the management process. Creating the right environment appears to significantly contribute towards achieving the project objectives.

REFERENCES

BS 5750, *Quality systems*, British Standards Institution

BS EN ISO 9000, *Quality systems* [formerly BS 5750], British Standards Institution

CIRIA Report 85, Target and cost-reimbursable contracts (1985)

CIRIA Special Publication 125, *Control of risk: a guide to the systematic management of risk from construction* (1996)

CIRIA Special Publication 129, *Value management in construction: a client's guide* (1996)

CIRIA Special Publication 132, *Quality management in construction – survey of experiences with BS 5750. Report of key findings* (1996)

CIRIA Special Publication 150, *Selecting contractors by value* (1998)

Construction Industry Institute, *Incentives in construction contracts*, Source Document 8 (1986)

Dorter J B, "*Partnering – think it through*" in *Arbitration; The Journal of the Chartered Institute of Arbitrators*, Volume 63: **3** (1997)

Egan J, *Rethinking construction*, DETR (1998)

European Institute of Advanced Project and Contract Management (Epci), *Complex capital projects and life cycle perspectives*. Working Paper (1995)

HM Government White Paper, *Modern local government – in touch with the people*, The Stationary Office (1998)

HM Treasury Construction Procurement Guides:
Guide 4, Teamworking, partnering and incentives
Guide 5, Procurement strategies
Office of Government Commerce (1999)

IChem.E, *Target cost contract form* ["*Green book*"]

Latham M, *Constructing the team*, The Stationary Office (1994)

Nolan Report, *Conduct in public service*, The Stationary Office (1997)

FURTHER READING

Chapman CB and Ward SC, "The efficient allocation of risk in contracts" in
The International Journal of Management Science, Vol 22: **6**, pp 537–552 (1994)

Chapman CB and Ward SC, "Evaluating Fixed Price Incentive Contracts" in
The International Journal of Management Science, Vol 23: **1**, pp 49–62 (1995)

CIRIA Publication C508, *Guide to developing effective learning networks in
construction* (1999)

CIRIA Publication C534, *Civil engineering design and construct – a guide to
integrating design into the construction process* (2001)

CIRIA Special Publication 117, *Value by competition: a guide to the competitive
procurement of consultancy services for construction* (1996)

HM Government, *Final report of the Government/industry review of procurement
and contractual arrangements in the UK construction industry*, HMSO (1994)

HM Treasury Construction Procurement Guides:
Guide 1 Essential requirements for construction procurement
Guide 2 Value for money in construction procurement
Guide 3 Appointment of consultants and contractors
Guide 6 Financial aspects of projects
Office of Government Commerce (1997–1999)

Steckelmacher P J, *Incentivisation of the supply chain in Government construction
projects*, Report for HM Treasury, March 2000

University of Reading, *Trusting the team – the best practice guide to partnering in
construction* (1995)

APPENDICES

APPENDIX A – TOOLBOXES

TOOLBOX 1 – GLOSSARY OF TERMS

ALLIANCE CONTRACTING
A form of partnering where contractors and clients, either together or separately, can form working relationships to achieve agreed objectives. More specifically they are supplier-led or project-led, as described below.

Supplier alliance
A structure where two or more contractors enter into an *ad hoc* alliance contract on a particular project to provide a working relationship with the client acting as the principal.

Project alliance
A structure where the client can be one of the alliance parties. A board of stakeholders, acting as the principal, usually controls the alliance.

CONTRACT FORMS
ACE – Association of Consulting Engineers – Fee Scale (permits other forms of reimbursement)

GC Works – General Construction Works – Government Contracts

ICE 5th – Institution of Civil Engineers – 5th edition – Full Design – Remeasure (does not contain contractor design unless amended)

ICE 6th – Institution of Civil Engineers – 6th edition – Design and Build (permits some contractor design, but is not a D&B for of contract per se)

IChemE "*Green book*" – Institution of Chemical Engineers – Cost Reimbursable (is a lump sum form as opposed to fixed cost)

IChemE "*Red book*" – Institution of Chemical Engineers – Fixed Cost

MF/1 – Mechanical and Electrical Works – Joint Committee of Institution of Mechanical Engineers and Institution of Electrical Engineers

NEC/ECC – Engineering and Construction Contract

NEC/PGC – New Engineering Contract

EXTENDED ARM CONTRACT (EAC)
An EAC is an agreement which creates a pseudo-partnering arrangement between a client and a contractor. The contractor will manage a large number of separate projects over an extended period of time. A form of management contracting ie employing a contractor to manage the works for a fee and employ discrete trades contractors to carry out the works. The advantages of this approach are that it is a fast track procurement route with the design and construction expertise already built-in, avoids the risk of compiling a single large package and spreads the construction risks, project by project.

FRAMEWORK AGREEMENT

An agreement promoted by the client that is held with a single or a small number of contractors, for a long-term business relationship for mutual benefit. Generally they have pre-qualified through an open selection process. They operate under agreed terms and conditions and tender for works as they arise. The benefit to the client is the reduced time taken in the tendering process. The commercial opportunity to the contractors is the restricted number of competitors.

INCENTIVES

Rewards, usually financial (or sanctions) which when applied within the framework of a contract will motivate all parties to improve their performance towards common project objectives.

LIQUIDATED and ASCERTAINED DAMAGES (LADs)

A genuine pre-estimate of the potential loss to the client, in the event of delay.

NEGOTIATED TENDERING PROCESS

Following receipt of tender proposals, the preferred bidder is drawn into a negotiation to fix the price and other terms.

OPEN BOOK ACCOUNTING

A process combining the scrutiny of project costs with the payment mechanism. The project accounts are maintained by the contractor and are open to full inspection and audit by the client. Payments made by the client to the contractor are calculated on the basis of the audited project costs, including contractor's profit attributable to the project.

PARTNERING

An arrangement whereby parties align their objectives on a project-by-project basis to achieve mutual benefits. This can be in the form of oral or written agreement. Some partnering agreements do override the legal contract between the parties and are legally binding. The base contract only comes into play in the event that the partnering agreement is breached.

TARGET CONTRACTS

A contract where there is an agreed apportionment of commercial and financial risks that may result in benefits to both parties to the contract if the delivery programme is improved upon. They can include risk and reward, not only for the programme, but also to performance standards or actual costs.

TURNKEY CONTRACT

A contract where the client hands over the whole procurement and construction process to a contractor who is responsible for the design, tendering, construction and commissioning of a project. The completed project is handed over to the client.

VALUE ENGINEERING

The process carried out by a design team and the contractor to review the design options in the context of whole-life costs to determine the best value solution for the client.

TOOLBOX 2 – WHEN SHOULD INCENTIVES BE CONSIDERED?

Where clients are intending to promote a scheme, then they should consider the achievement of time, cost and quality objectives in the context of the complexity of the project. Generally, where substantial construction risks are foreseen or are likely to exist, then they should be adequately assessed prior to considering incentivisation. To prevent the misuse of incentives, the following list of questions should be answered prior to the tendering process (see box).

> - Can all known risks be precisely quantified in terms of potential delay and cost?
> - If not, how will unforeseen risks be handled during the construction phase?
> - Can the management of these risks be aligned with the project objectives and are they compatible with the client/designers/contractor roles and responsibilities?
> - Will the intended form of contract be suitable to properly apply the incentivisation scheme?
> - How will the tendering process handle the contract strategy?
> - Will the client's procurement rules allow the development of a partnering arrangement which supports the success of an incentive scheme with the contractor?

The increasing use of Public/Private Partnerships (PPP) under the Private Finance Initiative (PFI) to construct major infrastructure projects has seen the development of a negotiated procurement procedure. In summary, following the technical and financial evaluation of prospective bidders, the short-listed bidders submit proposals which quantify and cost the risks. Following further negotiations a preferred bidder is selected. The contract defines who is responsible for managing each risk and specifies the contractor's output in performance terms over the period of the contract, usually called the concession period. This approach provides a framework for applying an incentivisation process and raises a series of further questions to be considered to obtain the full benefits of incentive contracts when applied to the traditional procurement route.

> - Has full support for the application of an incentive scheme been obtained from all levels within each party?
> - If not, how will the culture change needed to successfully apply the scheme be handled?
> - What criteria have been set out to define the success of the project?

TOOLBOX 3 – APPLYING INCENTIVE SCHEMES

Provided there have been satisfactory responses to the questions in Toolbox 2, the promoter of the scheme needs to define the outputs required in precise terms. This should not only include the tangible aspects of the construction requirements, but also the management and behavioural aspects of how the contract is to operate, what measures are to be used to assess performance and to obtain feedback upon completion. This can be broken down into three stages.

1. Pre-contract

- Has a market sounding exercise been undertaken to obtain feedback from industry as to what sort of scheme could be applied?
- Are there any special considerations which need to be promoted for the particular construction sector?
- What form of contract is best suited to promote the project and handle the incentivisation process?
- Have the hard and soft issues described in Section 6.2.1 been considered and adequate responses obtained?
- Are the contract documents concise, unambiguous and give a clear picture of the division of responsibilities and risks between the parties?
- Prior to contract award, is the client satisfied that the contractor's proposals agrees with their own views on the problem areas and scope of works?

2. Operation of the contract

- Broadly, are the administrative arrangements helping or hindering the management of the scheme?
- If they are hindering, what steps will be taken to improve the situation?
- Are the personnel from all parties to the contract fully aware of their roles and responsibilities and can any improvements be made?
- Is there an ongoing review of performance and what indicators are being used?

3. Post-contract

- Has the client prepared a questionnaire to allow all key personnel, including directors, to give their views on how well the scheme has been operated?
- What feedback process has each party instigated to promote the issues within their organisations and how will these be reviewed and any improvements implemented?
- If, within the public sector, the use of incentive schemes is considered to be successful how will central and local government authorities promote the lessons learnt and develop further opportunities to use incentivisation as a means of gaining better performance?

APPENDIX B – CONSTRUCTION PRODUCTIVITY NETWORK (CPN) WORKSHOP

LIST OF DELEGATES IN ATTENDANCE

Name	Position	Organisation	Status
Mr D Ajzenkol	Senior engineer	Charles Haswell & Partners	Delegate
Mr J Banner	Senior contracts manager	DERA	Delegate
Mr D J Barrow	Business development manager	Brown & Root Ltd	Delegate
Mr V Bevan	Estimator/ office manager	South Wales Police Headquarters	Delegate
Mr R Bishop	Design manager	Laing Technology Group Ltd	Speaker
Mr G Bradburn	Group commercial director	Drake & Scull Ltd	Delegate
Dr J C Broome	Research fellow	The University of Birmingham	Delegate
Mr M S C Brown	Construction management director	Entec UK Ltd	Delegate
Mr B Butterworth	–	Taylor Woodrow Construction Ltd	Speaker
Mr P Cooling	Project manager	Poole Borough Council	Delegate
Mr M Crick	–	Fitzpatrick Contractors Ltd	Delegate
Mr A Darby	Divisional director	Mott MacDonald Group Ltd	Delegate
Mr K Davis	–	Cyril Sweett Ltd	Delegate
Mr J R Dawson	Contracts manager	Union Railways Ltd	Delegate
Mr C Fuller	Principal consultant	Mouchel Consulting Ltd	Speaker
Mr T Gorman	–	Amey Rail Ltd	Chairman
Mr G Gray	Consultant	CIRIA	CIRIA staff
Mr C Hall	Principal	Weeks Technical Services Plc	Delegate
Mr J Hazelton	–	Thames Water Utilities Ltd	Delegate
Mr A Head	Chartered quantity surveyor	Anthony Collins Solicitors	Delegate
Mr G Hill	Chief civil engineer	Alstom T & D Systems Ltd	Delegate
Mr J Hughes	Construction unit head	South Wales Police Headquarters	Delegate
Mr I N L Jones	–	Water Service	Delegate
Mr I Kitchen	Director	Entec UK Ltd	Delegate
Mr P S Knight	–	London Underground Ltd	Speaker
Mrs H Kraus	–	–	Delegate
Mr J D Lloyd	Senior consultant	Edwards Project Management Ltd	Delegate
Mr C N P MacKenzie	Project director	AMEC Civil Engineering Ltd	Delegate
Ms P Moore	Senior quantity surveyor	Turner & Townsend Group	Delegate
Mr D Nicolini	–	The Tavistock Institute	Delegate
Mr K M Pagan	Director	Crispin & Borst Group Services Ltd	Delegate
Mr J A Park	Director	Stride Treglown Group plc	Delegate
Mr D Richmond-Coggan	Senior project manager	Mouchel Consulting Ltd	Speaker
Mr D Rogers	Contracts manager	Southern Water Services Ltd	Speaker
Mr J Seckerson	Managing surveyor	Lovell Construction Ltd	Delegate
Mr B Shepherd	Managing surveyor	Lovell Construction Ltd	Delegate
Mr A Stevens	–	Bovis Lend Lease	Delegate
Mr P Strickland	Contracts manager	Lovell Construction Ltd	Delegate
Mr A Wild	Senior lecturer	University of Central England	Delegate
Mr C M Winkler	Consultant	Miller Civil Engineering	Delegate
Ms K Wood	Project manager	Arup Project Management	Delegate
Mr M Young	Innovations manager	Balfour Beatty Ltd	Delegate

Construction
ProductivityNetwork

MINUTES OF CPN WORKSHOP

Construction contract incentive schemes: lessons from experience

Members' report E0120a

6 STOREY'S GATE
WESTMINSTER
LONDON SWIP 3AU

Telephone: (+44) (0)20 7222 8891
Fax (44) (0)20 7222 1708
Email: enquiries@cpn.org.uk
Web:www.ciria.org.uk

Report of a workshop held at the Royal Institution of Chartered Surveyors, London on 10 May 2000

Principal speakers	David Richmond-Coggan	L G Mouchel & Partners Ltd.
	Colin Fuller	L G Mouchel & Partners Ltd.
	Peter Knight	London Underground
	Don Rogers	Southern Water
	Barry Butterworth	Taylor Woodrow Construction
	Roy Bishop	Laing
Chairman	Tim Gorman	Amey Rail Ltd.

BACKGROUND AND OBJECTIVES OF THE WORKSHOP

Clients are increasingly turning to contract incentive schemes to improve performance and reward effort to promote a successful project outcome, allowing the contractor to benefit in some way. However, these schemes can be difficult to set up, the main difficulties being benchmarking, clarity of potential benefit, and motivation of participants at all levels. At this workshop the results of a recent CIRIA research project were discussed, including clients', consultants' and contractors' experiences of incentivised contracts.

DAVID RICHMOND-COGGAN – DIRECTOR, GOVERNMENT SERVICES, MOUCHEL

David is head of Mouchel PFI Unit, currently advising on a major accommodation PFI project for the MoD. He has 15 years' experience of working with public sector organisations to raise the quality of procurement and has sat on the CIRIA steering group for a research project on construction risk.

COLIN FULLER – PRINCIPAL CONSULTANT, MOUCHEL

Colin has extensive experience of business consultancy and senior management, including an operational audit of capital revenue expenditures in major engineering infrastructures for London Underground and VFM studies of development of best practice aimed at improving the financial performance of engineering and infrastructure.

Introduction

Incentivisation can be achieved in practice, according to those who have tried it.

They cite:
- improved performance
- significant benefits to <u>all</u> parties
- improved certainty of future programme planning
- improved projects to time, cost and quality.

Thanks to the steering group and case study interviewees, much has been learnt.

Overcoming the barriers

Two aspects from this part of the report were highlighted.

i) <u>Transparently identifying risks</u>
- has been shown by research to be a key determinant of successful outcome
- entails management action and communication
- assists in avoiding a 'claims culture'
- reduces risk of lowest bid avoiding risk issues
- improves accountability
- assists moving to a more risk-aware culture.

The CPN operates as part of the Construction Best Practice Programme. It is managed by CIRIA and supported by the DETR and the CIB

CIRIA C554

117

ii) <u>Tensions of competition vs co-operation</u>
- construction procurement historically based on extremely competition
- perceived demand to satisfy accountability
- producing low-cost/low-quality service
- co-operation may be seen as more 'grown-up' way
- how to avoid loss of demonstrable fairness
- trust is not enough
- different skill sets are required.

Why the document looks like it does
In researching existing experience of contract incentive schemes, it was found that literature was disappointingly limited, consisting mainly of
- CIRIA reports, 'Rethinking Construction' and 'Constructing the Team'
- Epci, CII and CIA Journal (Porter JB).

The second part of the research consisted of a number of case studies:
- 20 studies across a variety of industries
- building, power, transport, water
- a variety of underlying contract documents
- a variety of incentivisation mechanisms
- items covering all the incentive categories.

Evaluation
Experience was interpreted in terms of:

<u>what, why, how:</u>
- concepts
- categories
- benefits
- sectors

<u>environment, process, barriers:</u>
- objectives, targets, motivation, strategies
- decisions, agreement, documentation, management
- culture, relation to contract and procurement

<u>implementation</u> (and checklists in toolbox):
- when to consider
- step by step.

Lessons learnt
The case study lay-out includes views from all parties and suggested improvements, and throughout the text presents relevant illustrations with reference to specific case studies, as well as general reference, in order to break up discussion and add general interest. Flow charts and tables and summaries are also included.

The way forward
- Improved outcomes and measurable benefits have been identified.
- There are problems to be overcome.
- Investment in time and effort is required.
- Robust audit trails through negotiations are indicated.
- Some rewards are soft, not all are financial.
- Should schemes be 'sector based'?
- Culture change – partnering approach?

It is increasingly recognised that new, 'softer' communication and interpersonal skills are needed to promote communication, motivation, and joint problem-solving, thus avoiding the old adversarial approach and working towards a more stable procurement environment. It is not suggested that there is no competition, but rather competition which contains a negotiation element within it.

Conclusion
Contract incentive schemes are not necessarily an easy or comfortable option and need genuine commitment from all concerned in 'a different way of doing business'. *The good news from those who have tried it: it can be done!*

PETER KNIGHT – PROCUREMENT MANAGER, INFRACO BCV LTD, LONDON UNDERGROUND
A mechanical engineer with an MBA from Essex University, Peter has spent 25 years in the construction industry and has experience with Docklands Light Railway, New York Subway and London Underground Ltd. Infraco BCV is one of the three infrastructure companies soon to be in the private sector under the PPP contract incentive.

The client's view
The opportunities to be realised by applying incentives (and rewards) in the London Underground environment are huge, but with the following conditions.

<u>Trust is in the relationship:</u>
- trust is not the only ingredient, but it is critical that both sides understand
  - each other's objectives
  - the improvement promoted by the incentive
  - agree that the improvement is achievable
- commitment to the project and the incentivisation
- incentives may not be realised if things 'go wrong', when the strength of the relationship will be tested.

<u>There are greater opportunities in 'rolling programmes':</u>
- incentives against one-off projects can be very beneficial
- lessons learnt (and their cumulative benefits) from programmes of work
- added incentives to the supplier and to the client
- providing infill when a task really does finish early.

London Underground's 400-plus escalators, requiring constant attention, provide an obvious target for a rolling programme combined with an incentive scheme. An understanding of the base cost is a key factor; the client must be able to weigh up cost and return.

Workshops for potential contractors and key stake holders are held prior to the project to focus on the impact of the work and what can be achieved, such as replacing an escalator in sixteen weeks. Potential obstacles, such as hot working restrictions, soon become clear and these, as well as 'softer' issues, can then be addressed, promoting trust and team-working. Again, this works best with a rolling programme.

Achieved benefit justifies the incentive:
- the client must understand the added-value that an incentive can bring
- the client business as a whole must understand 'why incentivise'
- a simple formula is needed that does not encourage … argument
- give some thought about whom to incentivise, and how.

Improvements to realise the incentives are 'SMART':
- traditional use of time and cost targets; are they achievable?
- trust element when agreeing to the targets; is there an open book?
- contract strategy helps to stay focused
- risk management is a key component
- a SMART approach to achieve the desired outcome.

Conclusion

Involvement in the CIRIA study:
- confirmed there are opportunities for contract incentive schemes
- confirmed the need for 'extraordinary drivers'
- showed incentive schemes are best applied to rolling programmes but may also be effective in one-off projects
- showed that building trust is a key ingredient for success
- highlighted the importance of local accountability and problem resolution taking place at the lowest possible level.

DON ROGERS – CONTRACTS MANAGER, SOUTHERN WATER

During a career concentrated on project management and contract administration, including five years as deputy managing director of an engineering consultancy, Don has 30 years' experience in the water industry. He is currently responsible for the procurement of Southern Water's contracts and is on the ICE standing list of adjudicators.

Incentives versus disincentives

Why introduce incentive schemes?
- late completion of contracts
- cost overruns – 127% variance between tender and outturn cost
- defects
- adversarial attitudes
- duplication of effort.

Procurement strategy:
- establish long-term relationships with contractors
- align contracts terms more closely to business needs
- monitor achievement through audits
- provide incentives for performance.

Objectives of incentive schemes:
- move focus from short term financial gains to long term benefits
- provide real value for money
- attack costs rather than margins
- provide mutual benefit for client and contractor.

Characteristics of an effective scheme are:
- related to the prime purpose of the contract;
- balanced and aligned to aspirations
- simple to operate
- understood by those who will operate it
- documented
- seen to be creditable
- seen to be achievable
- focused on the long term.

Balancing the incentives entails:
- aligning aspirations of client and contractor, while taking into account
- commercial objectives of the contractor:
  - profit
  - repeat business
  - industry recognition;
- stated requirements of the client:
  - time
  - quality
  - cost.

The timing of incentives:
- pre-tender stage:
  - by client within terms of the tender document
- during tender:
  - by contractor within his offer
- post-contract:
  - by either party and introduced by agreement.

Continuous improvement is by means of incentivised targets:
- continually improving 'target scores'
- 'trigger scores' instigate defined procedures that will investigate the reason for under-performance
- the contractor is incentivised to achieve above 'remedy score'.

Financial
The percentage by which the contractor's application for payment value differs from the audited value has been significantly reduced over three years:
- target: 1.0% to 0%
- trigger: 3.5% to 2.5%
- remedy: 6.0% to 5.0%.

Safety
The number of accidents divided by the total number of work sites has likewise been reduced over the same three years:
- target: 0%
- trigger: 2% to 1%
- remedy: 3% to 2%.

Damage to utility plant
The number of instances of damage caused, divided by the number of excavations was also reduced:
- target: 2% to 0%
- trigger: 8% to 6%
  remedy: 12% to 10%.

Contractor selection considers the following factors:
- people
- culture
- team-work
- co-operation.

Client behaviour:
- needs actively to create a relationship based on trust
- maintains open and frank dialogue
- shows understanding of what enables the contractor to be efficient and effective
- is prepared to review own procedures and working practices
- understands the contractor's need to make a profit.

It is important that contractors' margins are maintained by driving down cost rather than profit. The amount saved in cost may then be offered back to the contractor as a bonus for meeting the deadline for completion.

BARRY BUTTERWORTH – PRODUCTION DIRECTOR, BUILDING DIVISION, TAYLOR WOODROW CONSTRUCTION (SOUTHERN) LTD.
Barry is a production director for Taylor Woodrow Construction, responsible for multi-million pound building projects in the south-east of England. Current projects include a residential scheme at St. Katharine's Dock, the Montevetro building at Battersea, a large office development in Fleet Street, and a PFI health project at Wycombe and Amersham hospitals.

Much of the CIRIA report is concerned with the public sector and includes a quote from Egan: *'clients need better value from the project and construction companies need reasonable profits to assure the long-term future.'*

In traditional time-cost-quality tendering:
- quality is a given specification
- time is the shortest programme
- cost is the lowest price.

This can lead to several problems:
- a disincentive to the contractor who will try to exploit to make a return
- in procurement there is no incentive to put forward alternative ideas
- the client's cost consultant cannot adjudicate if he cannot make accurate comparisons
- confrontation with the design team
- client's exploitation of a contractor's innovative idea by inviting more tenders.

As contractors' main concern is risk, affecting time and cost. This should be recognised by inviting the contractor to the table as early as possible. If there is good support by subcontractors, this will help develop trust within the team.

For example, one housing construction (luxury flats) project consisted of five separately negotiated phases. Movement in the housing market demanded early completion, which was jeopardised by risks presented by labour resources, planning conditions, environmental health regulations and wrong sequencing.

The flexibility created by the separate phases enabled client, contractor and subcontractors to reverse phases 4 and 5, optimising use of time, which in turn resulted in completion fourteen weeks ahead of time. Thus a 'way out' of problems was negotiated, without the need for new contracts.

ROY BISHOP – LAING
Roy is a chartered architect and was a CIRIA research manager for two years, investigating 'construction contract incentive schemes' and 'managing the design process in civil engineering design-and-build'. He joined Laing in 1997 as a design manager and currently heads Laing's 'knowledge transfer and best practice' programme.

Identification of objectives
It is fundamental that, from the beginning of the process, there is clarity of purpose, mutually sought wherever possible. Clear aims and objectives are essential to making incentives work.

Equally important is the investment of time to make sure the real issues behind the incentive (which may not be those first stated) are understood.

It is also useful for both parties initially to define the meaning of the incentives, the understanding of which is often an issue in response to bespoke procurement opportunities.

The report provides useful guidance in putting incentive schemes into three categories:
- pre-planned
- responsive
- reactive.

A thorough investigation of these headings and their meanings is recommended, as they will clarify the roles, responsibilities and opportunities of the parties.

Incentive mechanisms
'What gets measured gets improved'. It is essential to understand what measurements can be applied to project performance, as there is a great need to couple the contract strategy to the project delivery. These measures and mechanisms should be transparent and readily understood by all.

If necessary, these measures and mechanisms may even be placed outside the contract itself as measures for all the parties to understand their own performance, rather than as terms for argument or grievance.

Organisational interfaces must be properly addressed, and trust is vital in this area. However, risk apportionment and clarity of understanding are the key items underlying that trust.

Personal responsibilities are also critical. Culture is an important factor and the aim should be to bring the right people to the project and to 'push down' the level of authority as far as possible.

Communication
Even with board-level buy-in to this concept, board, management and 'coal face' levels may be speaking different languages. Considerable investment is necessary for consistent, clear and repeated communications to be 'translated down' into meaningful statements that can be understood at all levels.

The weakest point in the chain is at the site level and it is important that the at all levels of the project communication should be swift, clear and unambiguous, using workshops, launch meetings, even project web pages.

Feedback is also an important factor, and when it becomes necessary to address issues of performance, it is recommended that this be attempted initially at the lowest level; if issues are allowed to rise up the organisation, the opportunity for early resolution, and for removing obstacles to effective performance, is lost.

DISCUSSION

Q: Will new skills be required when considering softer benefits, customer loyalty etc?

A: New skills will need to be developed with regard to culture and communication.

Q: In the experience from the CIRIA report's case studies of incentive schemes or incentivising alliance, should this involve the entire team or single parties and should the incentive be financial? Are there any developments regarding the pre-tender/post-tender issue in the public sector?

A: When the conventional procedure of competitive tender is adapted to include an element of negotiation, it is possible, especially in the public sector, that the standard procurement processes cannot be met. Hopefully, the CIRIA report's recommendations, based on experience of incentive schemes, will help to show the way ahead and give confidence, including that in the public sector.

The MoD, for example, in a move towards employing prime contractors within its construction procurement, are promoting the appointment at an early stage of a single voice

for the project in the form of a contractor, to manage the project and deliver its various elements.

The problems of the design component of the team have been highlighted; as designers work on a different basis from subcontractors, the question of their share in the benefit also needs addressing.

Q: What are the gains of an incentive scheme encouraging those aspects which the contract itself is designed to achieve?

A: This is a grey area. Incentives may help to deliver contractually agreed provisions as well as bringing benefit. Bonuses are often given to achieve what a contract requires in meeting deadlines.

Q: How can fragmentation of subcontracting, such as the indirect employment of labour, and the lack of rational use of labour, be overcome by offering incentives?

A: Paying labour by the day does not act as incentive to achieve daily objectives. Objectives need to be delegated to subcontractor level to incentivise the 'coal face'.

Q: It has been said that in London the bulk of labour lacks incentive and, furthermore, the use of hourly rates acts as a disincentive to prompt execution. Yet clients may insist that incentivisation extends down to the workforce. How can this be addressed?

A: This feeds back to the concept that the mechanics under discussion may not apply at the vital workforce level. There appears to be reliance on the industry's structure to ensure performance by all strata but this is prone to failure, often due to vulnerability to disruption from below, creating weakness in long-term delivery.

The industry as a whole must take account of those situations where there is little communication or benefit, whether or not a formal incentive scheme is in place.

Employing the workforce is a procurement issue and an integral part of supply chain management and as such needs close examination. This requires an initiative from the procurer, who is the client, and professionals should talk to clients to make them aware of this.

Incentive scheme deals tend to be made at high level, making them more difficult to implement with subcontractors. Other contributors to failure may be the history of the construction industry with its reliance on casual labour, and lack of apprenticeship schemes: a deep-seated problem which has developed over the last 20 years and cannot be undone quickly.

However, it should be possible to devise an incentive scheme aimed directly at a subcontractor's workforce. A system of responsibility and reward may also help to provide motivation.

Q: Where there is a substantial interface, such as with rail and road networks which is subject to consents, both client and contractor can underestimate the risk attached to obtaining consents. Has the CIRIA report considered examples of projects going awry due to difficulties with risk assessment?

A: This is one of the key issues of the report which describes three approaches: pre-planned, reactive (a valuable application of the incentive scheme) and proactive. Well-defined tender schemes may become liquidated damages in reverse and this is where tailoring to requirements comes into its own and can be very successful.

This requires not only a combined evaluation but transparency about outcome, establishing a firm basis from which to renegotiate. Such a secure agreed understanding of risk to be shared and borne could not be accommodated by the old system of competitive tendering.

Q: It has been suggested that the fundamentals of risk assessment are often not understood and therefore need expert consideration. How does the report address the issue of risk management?

A: It does appears that, except on major projects where professional risk experts are employed, engineers are often expected to act as project/risk/financial managers without the appropriate skills, and many contracts have gone awry because risk was not investigated in detail or at a sufficiently early stage.

However, a new culture does seem to be developing, and with it a body of expertise and special interest groups concerned with addressing risk by designing toolkits and distributing 'best practice' material.

Q: In risk management, the question 'Who owns the risk?' is fundamental. How is this dealt with?

A: Risk ownership has been addressed in the report. Risk needs to be managed and monitored and funds allocated.

Q: How can the base of incentivisation be broadened to take account of local issues, for instance in the promoting of local employment and other benefits to local economies?

A: Non-cost/time benefits and other soft issues are hard to evaluate. There needs to be clarity of purpose which is communicated to those on the ground, where benefits may not be obvious.

Social impact measures need to be assessed as environmental issues are addressed, while both parties need to appreciate each other's business and its respective drivers and aim to align these.

Q: How effective were clients in communicating their drivers and challenging objectives, and conveying these to their team?

A: Some schemes were more successful than others and it seems that those without a clear contracting strategy did less well; informal incentive schemes based on personal negotiation were the most successful. A client's ability to communicate effectively with the contractor appears to be the key determinant.

Q: How can incentive schemes be structured to improve and sustain contractors' performance long-term, especially if benefits have few tangible aspects?

A: This has not been part of the study which takes a point-of-time view of a number of specific construction projects, considering the environment and studying the history of relationships. However, a number of points arise.

Projects
These tend to fall into three categories.
1. Large construction projects with very long timescales.
2. Programmes of projects where incentive schemes allow demonstrable and continuous improvement from project to project (incentive not lost).
3. Maintenance/long-term contracts.

In the third category, incentive schemes also rely on continuous improvements which are measurable to ensure continuation. The MoD aims at employing similar mechanisms in prime maintenance contracts.

Softer aspects of incentives may include:
· benefits of long-term relationships
· stable supply chains
· learning from experience.

The resulting efficiency can lead to measurable savings. Long-term relationships should therefore not be underestimated and are always to be recommended, even if, with time, benefits increase more slowly.

Q: Surely a good incentive scheme becomes a driver for trust as both parties have a common interest. Is there a methodology in allocating incentives which will lead to gains?

A: This issue is addressed in the report, which includes a section on implementation for guidance. However, some of the issues are difficult to understand and specifics and generalities need to be addressed separately.

Central to a scheme's success is local decision-making and delegating responsibility to those who know what it means and how to achieve it.

More feedback is needed; at present samples are too small to be representative. This area should be revisited in about two years' time.

Q: What can be done about the erratic behaviour of a client who is disorganised or becomes progressively more so as a project progresses?

A: One way is to tailor the scheme to suit the situation as project needs change, including, if necessary, a change of contract or change of project manager. 'One-off' clients in particular will need guidance to avoid becoming disorganised, as they lack experience of how projects - and agendas - can alter.

Longer-term contracts tend to favour incentive schemes, with contractors more ready to help manage change and to recognise skills shortages. Examples can of this can be seen in areas of privatisation, where partnering and incentivisation have proved very successful.

Incentive schemes will only continue to be adopted if they are seen to be successful, and they need drive and resilience. Much depends on the personalities of those charged with their implementation.

Q: Many of the issues discussed have been addressed in the USA. Have any conclusions been taken into account in the CIRIA report?

A: Some of the US literature was studied and found not significantly different. However, some considerable differences, such as environment, have to be recognised, and although there may be pointers, they do not provide the necessary relevance for clients facing procurement problems in this country.

CHAIRMAN'S SUMMARY – KEY POINTS

- There are different stages at which incentive schemes may be applied:
  - tender stage
  - responsive
  - reactive.
- There is a need for expert risk assessment.
- New skills are required in projects management.
- Trust is a key feature of incentive schemes.
- A rolling programme promotes continuous improvement, which provides incentive to the contractor and encourages building on improvements achieved.
- The question of incentivising the workforce needs to be addressed and those who influence the outcome must be identified.
- Are all parties to the contract (client, contractor, other consultants) part of the incentive scheme and have potential blockers been identified?
- Evaluation of soft benefits beyond the features of the contract needs to be considered.